EVOLUTION AND CONSTITUTION

THEORY AND DECISION LIBRARY

General Editors: W. Leinfellner (*Vienna*) and G. Eberlein (*Munich*)

Series A: Philosophy and Methodology of the Social Sciences

Series B: Mathematical and Statistical Methods

Series C: Game Theory, Mathematical Programming and Operations Research

SERIES A: PHILOSOPHY AND METHODOLOGY
OF THE SOCIAL SCIENCES

VOLUME 37

Scope: This series deals with the foundations, the general methodology and the criteria, goals and purpose of the social sciences. The emphasis in the Series A will be on well-argued, thoroughly analytical rather than advanced mathematical treatments. In this context, particular attention will be paid to game and decision theory and general philosophical topics from mathematics, psychology and economics, such as game theory, voting and welfare theory, with applications to political science, sociology, law and ethics.

The titles published in this series are listed at the end of this volume.

EVOLUTION AND CONSTITUTION

The Evolutionary Selfconstruction of Law

by

Erhard Oeser

University of Vienna, Austria

KLUWER ACADEMIC PUBLISHERS

DORDRECHT / BOSTON / LONDON

A C.I.P. Catalogue record for this book is available from the Library of Congress.

ISBN 978-90-481-6503-2

Published by Kluwer Academic Publishers,
P.O. Box 17, 3300 AA Dordrecht, The Netherlands.

Sold and distributed in North, Central and South America
by Kluwer Academic Publishers,
101 Philip Drive, Norwell, MA 02061, U.S.A.

In all other countries, sold and distributed
by Kluwer Academic Publishers,
P.O. Box 322, 3300 AH Dordrecht, The Netherlands.

Printed on acid-free paper

TABLE OF CONTENTS

PREFACE

Since more than two thousand years legal philosophy has dealt with the problem of the "trans-legal" or the so-called "just" law in an effort to establish the unalterable foundation of each legal system created in the history of humankind. At the end of this long road, however, one finds not the expected stable, universally recognized legal fundamentals, but rather the fact that neither legal science nor philosophy could hinder the "legal injustice" in our marked by world wars and political dictatorships century. One began to talk unmistakably about the "failure of legal philosophy" to provide a clear demarcation between justice and injustice.

However, this disappointment in legal philosophy as a theory of the "just law" proved to have a positive aspect as well. Thus, the textbooks on philosophy of law before 1945 were dedicated mainly to traditional philosophical disputes or to explanation and defence of empirical sociological or psychological research aiming to supersede the traditional legal philosophy. Nowadays the problems of interest are more content- and system-oriented ones. The questions of belonging to a particular philosophical school like neo-Kantianism, neo-Hegelianism, positivism, phenomenology, existentialism, or neo-scholastics have been replaced by questions concerning the relation between law and morality, the historicism of law, and the research and argumentation of what was earlier called natural law and law of reason. In the meantime it has become clear that juristic logic as the formal technique of logical argumentation could not adequately serve the contents-oriented and substantial reasoning of juristic decisions. A rational discourse referring not only to logics and theoretical reason but to practical reason as well (based on the model of profounder argumentation) is particularly necessary in cases which relate explicitly to the classical moral-philosophical concept of trans-legality. The traditional legal philosophy after Kant presented just law essentially as a non-historical phenomenon. On the contrary, nowadays one should presume the historical character of law as outlined by the historical school, i.e. legal philosophy as a theory of practical reason has to be preceded by a theory of historical reason

The changed view on trans-legality has proved to be of crucial importance. Thus the "above-law" legal-ethical principles are no more considered to be eternal and unalterable truths but rather dynamic principles revealing their real value in the process of argumentation of each particular case. Hence justice has to be fought for with rational arguments over and over again asserting it in the respective positive legal order.

Since Darwin this conception has always been related to evolutionary ideas. In the 19th century popular legal scholars like Ihering, legal historians like Morgan and – before these – legal ethnologists like Sumner Maine and Post, as well as the founders of comparative science of law Kohler and Bernhöft advocated the idea of evolution of law. Despite the abuse of this idea on behalf of social Darwinism (which explains the aversion to evolutionism of Dilthey and other philosophers oriented towards humanities), its positive gist remains nowadays more than ever sustained. Modern explanatory models of organic evolution have proved to be enough comprehensive to encompass the development of human being together with its cognitive and moral capabilities (Lorenz, Hayek, Mohr, Riedl, Vollmer). These models have fully justified the transferability of the evolutionary approach to complex social phenomena like economy and law.

In the present book the idea of evolution of law is not used as a superimposed biological metaphor. Rather an attempt is made to integrate it in the body of the traditional problems of legal philosophy. However one should not expect to see a description of legal-philosophical and legal-methodological systems in the vocabulary of evolution. As Habermas claimed, the application of evolutionary theory was not writing of history but rather practical discourse. If we also consider Alexy's statement that juristic argumentation is a "special case of the general practical discourse" then we'll be getting a direct connection between juristic argumentation theory and evolutionary approach to law in the sense of a developmental theory of practical reason. Hence the presented in the book idea of the evolution of law should not be understood (as in the case of other evolutionary conceptions) as a "theory of practical irrationality" referring to a "myth" (Luhmann) or to the "overestimation" of reason. True, the starting point here is also the basic evolutionary idea that not reason originates order (Hayek) but *vice versa* – the existence of order from the very beginning made possible the origination of reason. At the same time, considering the historical development of juristic methodology one can detect – despite all relapses, dead ends and unsolved problems – a "progress in the awareness of human freedom" (Kant, Hegel) stimulated by an increased legal self-determination of reason.

Hence one can speak of a kind of rehabilitation of law of reason in a new form consistent with the evolutionary approach: The prehistory of humans knows no natural state completely devoid of laws. Respectively, the existent in the history of mankind legal orders have not dropped from the sky but have rather developed in a rational argumentation process through internal self-construction. The mechanism of this evolutionary self-construction of law consists in the cognitive "opening" of the current legal system for particular facts of life and trans-legal practical evaluations and its respective normative "closing" in the ruling of each concrete case. The popular in hermeneutics "changing perspective" (Engisch) between legal norms and concrete cases functioning here as a developmental factor of law, is reasoned also on the basis of the theory of evolution. As all self-organising processes in nature the self-construction of law doesn't relate to individual instances but rather to a procedural mechanism. The latter however finds its "carrier" in the practical reason of each human being. Natural history becomes a history of mankind only through the intervention of reason. Hence everybody who believes in the progress of reason in the course of human history has to embrace evolutionism as well. The origin of humans from pre-given structures of order of the living nature should not be viewed as a self-abasement of man. On the contrary, in the spirit of Darwin the ascent of man to the highest level of organic evolution through his own efforts is an achievement to be valued.

The present book is the result of a course of lectures on juristic methodology and legal philosophy at the Open University in Hagen. The central idea of an evolutionary self-construction of law was born in a discussion with Friedrich von Hayek at a Darwin Symposium in 1981. The book would not have been possible without Prof. Dr. Peter Raisch's cooperation and advise who not only brought home to me the juristic way of thinking but had an ear for my evolutionary ideas as well. He gets my special and sincere thanks for our friendly discussions of long standing. My thanks go also to Ms Maria Dürr who unremittingly typed down all possible draw-ups and versions of the book.

Erhard Oeser

PREFACE TO THE ENGLISH EDITION

Since the appearance of the German edition of the book in 1990 evolutionary legal thought has developed not only in the sphere of legal philosophy and methodology (Goutier 1989) but also in private law (Behrends 1993), fiscal law, subsidy law, social insurance law (Helsper 1989), family law and environmental law (Gruter 1993). These developments have resulted in the conception that legal norms are nothing else but a higher form of a biologically founded behavioural pattern. On its behalf this conception has led to the appearance of a new province of legal science – behavioural legal science or legal ethology which provides both legal anthropology and legal history with a profounder (if not sufficient) explanatory foundation.

Something more, evolutionary legal thought (present in the English legal science since Darwin) could facilitate the integration process between case law on the one hand and normative law, on the other. The recent trends in legal theory of the German speaking region have clearly shown that even the most abstract norms and principles of an axiomatic-deductive legal system are nothing else but norm-hypotheses (Kriele) which have to be always adapted to the new situations in human development. Hence, the difference between normative law and case law could be only relative. The one who applies norms should know not only the legal code but also the respective cases. A judge confronted with a new situation has to cross the borders of the legal code thus creating law through his jurisdiction. One observes the same evolutionary learning effect in case law: A ruling becomes a judicial precedent only when the decision in a single case leads to the abstraction of a general scheme to be further applied in similar cases. In normative as well as in case law the jurists have to ensure the same prerequisites for the solution of unknown cases. The evolution of legal behaviour, which in both cases relates to the same biological foundation, has developed – in the course of human history – to a systematic self-construction of law. The latter transgresses the biological requirements for survival of the species, which reflects the fact that the legal principles construing the legal regimes of all states and peoples serve not only the biological but also the cultural existence of humans.

My thanks to those who through suggestions and discussions both in biology and law encouraged and helped my work on the German version. I would like to thank also all who showed interest in the book and criticised or corroborated the views presented in it. My special thanks go to the translator Maria Wuketits who – considering the terminological differences between biology and jurisprudence as well as between European and Anglo-American law – had to fulfil a difficult task.

Erhard Oeser

DESCRIPTION AND PURPOSE OF THE WORK

Evolutionary approach to law had been advocated already at the time of Darwin by English ethnologists of law like Sumner Maine and American anthropologists like Morgan. The present work is an attempt to apply evolutionary thought to the continental legal philosophy and juristic methodology. Although in the 19th century the idea of an evolution of law was present also on the continent it was burdened with reproaches of social Darwinism (proved nowadays as unreasonable). Grounds for the negotiability of evolutionary approach to complex social systems are provided by the latest research in evolutionary theory of cognition and evolutionary ethics as well as by the theory of complex systems in the sense of Friedrich von Hayek.

History of law is no longer perceived as a history of persons, events or norms but rather as an evolutionary self-construction, which is indeed founded upon the biology of human behaviour. Nevertheless it goes beyond the mere biological conditions for survival of the human species: The principles of law in all possible legal regimes take account not only of the biological but of the cultural existence of mankind as well. The legal-ethical principles themselves are not more understood as eternally valid truths but as norm hypotheses, which have to be adapted to the new situations in social development. This new approach could become the common foundation which to bridge the gap between case law on the one hand and normative law, on the other.

The book provides a profounder (conformable to the biological theory of evolution) explanatory basis not only for anthropology and history of law but also for juristic methodology. In addition it offers a practical model of juristic argumentation processes, illustrated by many examples. The book is addressed to philosophers of law, historians of law, theorists of science and social scientists as well to all those who are interested to overcome the gap between science and humanities.

0. PHILOSOPHY AND JURISPRUDENCE

Some philosophically minded jurisprudents advocate the opinion that the "textbooks in legal philosophy, though abundant, are as a whole controversial". What is more, legal philosophy could not be learned the way one learns penal law, law of succession or, for that matter, physics (Kaufmann, 1983, IX).

Therefore, it is little wonder that, while the German speaking countries have witnessed "an impressive amount of textbooks in legal philosophy" (Dreier, 1981, p. 33) in this century, "philosophy of law" as an academic subject has retreated further and further into the background, superimposed and superseded by other subjects like "legal theory" and "legal sociology" (Dreier, 1981, p. 17). Although the name "legal theory" could mislead one to believe that it is only a version of legal philosophy, it is nevertheless primarily a methodology of jurisprudence. As such it makes each traditional (one that deals with the concept of law and not with the legal acts as positive law does) legal philosophy impossible.

Not without reason Radbruch called the general legal doctrine generating the legal theory of legal positivism the "euthanasia of legal philosophy" (Radbruch, 1963, p. 114).

There have already been emphatic warnings against *legal science without law*, as Leonard Nelson (1949) calls it. (Schneider, 1972, pp. 108-110) The restriction to "normative-logical exercises" is compared to the "mission of medicine to dissect corpses and to classify disease, at the same time being not allowed to research the causes and effects of the disease itself" (Klenner, 1967, p. 1628).

The fact that the restriction of legal philosophy (actually acting as a self-cancellation) to one single aspect of its traditional themes (methodology) finds its equivalent in the "turn of philosophy toward philosophy of science", is clearly evident in the programmatic formulations of legal theory explicitly defining the latter as a legal "analytical philosophy of science" or as protojuridics" (Jahr, Maihofer, 1971).

What then remains for legal philosophy? Contemporary literature on legal philosophy and legal theory stimulates enough discussion on this problem by viewing legal philosophy as a "residual discipline".

According to Ryffel (1969) the term "residual discipline" reflects the situation of legal philosophy after the separation of those disciplines, which either made themselves independent like psychology, sociology, politology, etc., or became parts of legal science, e.g. legal theory or juridical methodology. Thus, legal philosophy is confronted only with "residual problems" which cannot be solved by the newly established disciplines, the latter having (with the exception of logics or juridical methodology) predominantly or solely empirical character. One way or another, these residual problems – being recurrent – should be dealt with. Ralf Dreier illustrated this quite clearly, in a variation of Horace:

Philosophiam expellas furca, tamen usque recurret – one could expel philosophy with the "dung-fork", it will nevertheless keep on returning (Dreier, 1981).

If, however, one brings the view of philosophy as a science about unsolved problems of particular disciplines together with the reproach against the "backwardness of legal philosophy" the term "residual discipline" is characterized only negatively.

Still, one could view the separation of single disciplines from philosophy positively as well. In this context, legal philosophy literature (Kaufmann, 1984, p. 7) refers to Heidegger's fundamental claim about the crucial turn philosophy had undergone already in the antiquity – a claim that is valid for both philosophy of science and history of science. According to Heidegger the crucial turn reflects the development of sciences within the horizon opened by philosophy. This development comprises both – the detachment of the sciences from philosophy and the establishment of their independence. The described process indicates the *Vollendung* of philosophy... In its history philosophy has tried sporadically and not sufficiently to present the ontology of what exists (nature, history, law, arts). Now all this is performed by the sciences (Heidegger, 1969, p. 63-64).

Under *Vollendung* of philosophy Heidegger seems to understand the termination of philosophy.[1] On the other hand this expression could be reflecting the ever-exacter specification of the fundamental problems of philosophy, which tend to appear in each science but prove to be unsolvable by the respective scientific methods. In such a case it would be inadequate to speak about the termination of legal philosophy. On the contrary, here we are witnessing a kind of a progress in the problem-oriented philosophical consciousness, bringing to light fundamental questions which normally go unnoticed as opposed to secondary empirical, anthropological, psychological, or sociological ones.

Yet, such an approach confronts legal philosophy with one necessary prerequisite, worded in the sense of Savigny: Those that are not convinced in the priority of legal philosophical reflections to the empirical research in law better leave it [legal philosophy] alone (Wesenberg, 1951, p. 50).

One could hardly deny the importance of empirical research – especially in the field of legal sociology. Nevertheless, the argument of the priority of

---

[1] Translator's note: The German phrase *Vollendung der Philosophie* could be ambiguously understood both as termination or fulfilment of philosophy.

legal philosophy as a fundamental theory of law is vital to the justification of its existence as an *apologia pro sua vita* – it is an immanent and coercive feature of philosophy, one that distinguishes it from all other sciences. Accordingly, one can make the very general claim that if there is something like legal philosophy then only in the sense of fundamental research of law. What legal philosophy reflects goes beyond juristic dogmatic, namely, to the pre-suppositions implicitly and non-reflected contained in the formulation of the so-called basic laws. The explication and argumentation of those more profound presuppositions comprise the intrinsic and ever-present task of legal philosophy, which no other discipline could ever perform. Neither philosophers nor legal scientists or practitioners have at any time contested the fundamental character of legal philosophy. "All jurists make use of legal philosophy, most of them, however, do not know it", stated Dreier (1981, p. 32) in reference to a remark by Popper (1973, XXV). Dreier's statement characterizes the contemporary situation of legal philosophy – as a term repressed, yet, as a subject-matter – unavoidable.

The re-emergence of legal philosophy – often proclaimed dead – is accompanied by a "rehabilitation of practical philosophy" (Riedel 1972-74) generally noticeable also in other spheres of human and social sciences, and the humanities. It is quite obvious as well that Dreier's statement about legal philosophy reflects no doctrine but rather a basic individual practical and philosophical attitude, revived by Engisch who using Fichte's famous words claimed. The choice of philosophy depends on the person who chooses; for a philosophical system is not something to discard or keep as one likes, rather it is animated by the soul of the person who possesses it (Engisch, 1971).

Most contemporary jurists share – also intuitively – the certainty that law and, moreover, positive law could be neither taught nor learned, neither established nor performed without the touch of legal philosophy. Still, this doesn't prescribe which of the existent legal philosophical systems from both the past and the present should be adopted.

Years ago Engisch outlined the situation most clearly: Whenever one glances at the numerous philosophical doctrines and systems emerging in the process of historical development and represented by prominent contemporary philosophers, one could not but despair when confronted with the necessity to choose among them... Sometimes philosophy is interested in the "being", in the reality of nature, sometimes again – in values. In the case of the latter it should not only contemplate and get to know them, it should – in the opinion of some – also create values (Nietzsche) even "change the world" (Marx). Correspondingly, the organon of philosophical cognition is not always and not exclusively the "ratio", the mind, or the reason but, in some cases, rather the intuition or the feeling (Engisch, 1971, pp. 2-3).

At the same time, Engisch addresses the problems of philosophy of science as well: The relation to the sciences is interpreted in different ways. Sometimes philosophy should provide the basic concepts and fundamentals of science, sometimes, however, it should constitute the corner-stone of the special scientific conclusions and synopses. And so forth.

Thus, legal philosophy is forced into the following situation: When conceiving law as a "natural law" or as a part of the "order of genesis", or as realized in some other way of the being, then it performs as legal ontology. If it contemplates law as a "cultural value" (Radbruch), then it appears as a branch of value philosophy. As a "general legal doctrine", respectively "legal theory", it is oriented towards "positive law" elaborating its general structure, its "basic concepts" and at the same time striving after more rigorous scientific character. Lately, as a "juridical logic", it makes even use of the methods of mathematical logic, the so-called "logistic" (e.g. Klug: *Juristische Logik* [Juristic Logic]). On the other hand, in view of the ancient but still very much alive relation between law and religion, legal philosophy moves closer to theology, being thus coerced into a "religious philosophy" of law (Radbruch). Sometimes legal knowledge makes use of "reason", sometimes of the "view of essence" – especially of "values", sometimes of "legal sense", providing it doesn't succumb itself to scepticism or "relativism".

By enumerating all these versions, Engisch presented still another, and rather broad concept of legal philosophy, traditionally common for the historical development of the discipline up to the 20th century and later, and also present to a bigger extent in its current interpretations. Such a broad view, however, is normally achieved at the expense of a comprehensive systematic presentation, which explains why most of the attempts in this aspect (if not solely or predominantly historical – e.g., Emge, del Vecchio, Verdross, Wolf, Sauter, Welzel, etc.), restrict themselves to the elaboration of the "basic concepts" or "main topics" (as e.g. Engisch himself) of legal philosophy.

0.1. *The Three Classical Dimensions of Legal Philosophy*

The view of legal philosophy as a part of philosophy – legitimised by a statement of such a prominent legal scholar as Gustav Radbruch (he begins his *Rechtsphilosophie* with the sentence "Legal philosophy is a part of philosophy") – reveals a classification maintained in all historically developed philosophical schools. Thus, since Aristotle, one has differentiated between two parts in traditional philosophy – theoretical and practical – subordinating logic as *organon* or "tool". Similarly, legal philosophy could be looked upon as comprised of a:

1. Methodological dimension,
2. An ontological dimension, and
3. An axiological dimension.

Regardless of the fact that these dimensions are differently evaluated with regard to their role and importance, as well as to the way they interact, studies of contemporary authors still show fundamental similarities. Thus, e.g. Engisch (1971) spoke about "three dimensions of legal philosophical research" which correspond to the above mentioned ones:

- Logical-methodological dimension, defined by juridical thinking in logical "judgements" in the form of propositions, the elements of which are juridical concepts.
- Ontological dimension interested in the "essence of law" and studying law "from within".
- The third dimension focuses not on law in itself, but rather on its appearance as a cultural phenomenon in its interaction with the various spheres of our existence: religion, morality, state power, etc.

In his "integrated legal theory" as a fundamental theory of law, Dreier (1981) distinguished between three aspects:

- Analytical aspect, referring to the logical and linguistic analysis of the basic concepts,
- Empirical aspect, reflected in legal-sociological and legal-psychological research, and
- Normative aspect, maintaining legal-ethical and legal-political problems.

Dreier's concept "integrative legal theory" is an attempt to re-integrate the disciplines that are drifting apart: On the one hand, the merely methodologically understood legal theory as an analytical theory, and on the other – the empiric-realistic theory and normative theory. Also Arthur Kaufmann (1984, p. 10) opposes such a separation between legal theory, legal sociology, and legal philosophy, emphasizing that it reduces the subject-matter of legal philosophy to not more than axiological structures. According to him a legal philosophy having to deal only with values would hang in the air. On the other hand, legal theory and legal sociology could not be exercised as juridical disciplines at all without the incorporation of legally normative criteria (Kaufmann, 1984, p. 10). Bearing in mind the structural intertwinement of "is" and "ought", Kaufmann claims the inseparability of the following three aspects:

- Logical aspect,
- Aspect of the "is", and
- Axiological aspect.

Even in those cases, where, as for example, in Hart's (1971) new analytical legal positivism, the formal-logical aspect is treated as the main aspect, and law and morality are strictly separated from one another, the other aspects are at least maintained as a minimum of metaphysics and as a practical argumentation.

0.1.1. *Methodological Dimension: Logic and Law*

Within juridical methodology, the methodological dimension nowadays is treated as a discipline in itself. At the same time, however, it forms an indispensable foundation for legal philosophy, oriented predominantly towards the problem of ethical values (axiological dimension). More than hundred years ago Ihering pointed to the fact that all legal philosophers that are not jurists themselves should bear in mind an aspect of practical morality; a purely ethical acknowledgement of a particular law is not possible without knowledge of the technique (Ihering 1852-63, p. 351).

It is, however, possible to keep both aspects – the formal-technical on the one hand and the one regarding the contents (ontological or axiological) – on the other, systematically separated in the following way:

- Juridical methodology deals with the "functioning of law in legal decision-making",
- Legal philosophy deals with the just law, possessing as an idea or a value an ontological-metaphysical as well as an ethical aspect.

The link between legal methodology and legal philosophy is to be found – like the General Legal Theory as a comprehensive or integrative discipline – in the logical-hierarchical structure of the legal system. That is, each particularly positive law should be – in its application in the sense of its logical foundation – traced back to a logically non-deducible basic norm, which nevertheless is ontologically, ethically, religiously (according to the twists and turns in the history of legal philosophy) founded.

This shows clearly that the relationship between logics and law is always based on the application of logics in the sense of an argumentation technique. Or, in a narrower sense, this relationship concerns itself only with the formal consistency of the legal system – a well-known fact.

According to Ihering the law of consistency refers not to contradictions due to the legislator, but rather to the contradictions of jurisprudence with itself. Thus, if one considers the way legislators or statute authors have implemented the codifications (available in legal history), it becomes clear enough that these codifications represent in no way a logical system according to the principle of consistency and completeness. In the Anglo-American *case law* the logical structuring of law is thoroughly denied (See Fikentscher 1975-77, p. 698).

On the contrary, modern representatives of juridical logics in continental law contemplate the task of logical structuring of the legal system as one of the main points of interest of jurisprudence. Thus Kelsen claimed that the legal propositions of jurisprudence, describing legal norms have to be logically connected. Thereby, one verifies the system of norms of positive law.

Klug maintained that the task of jurisprudence is to bring the system of valid positive law in the form of more exact juridical calculations.

Yet, even the most optimistic representatives (e.g. the above mentioned Klug) of the axiomatization theory realize that such a task can only be approached as an ideal claim that is to be resolved only approximately at best. Thus, Klug himself speaks of nothing more but a "quasi axiomatization" of law. Here, the difficulties stem mainly from the complexity of the sphere regulated by the legal system. The diversity of legally relevant human modes of behaviour, actions, reactions to actions, and reactions to reactions could hardly be grasped by an assessable system – on the one hand, it is impossible to describe their variety, on the other, one could not really foresee all future behaviour patterns.

Nevertheless, the logical dimension remains an indispensable aspect of any legal philosophy. No matter whether the latter is to be understood as ontology or ethics of law, one still needs to logically justify the connection between the legal norms of positive law, on the one hand, and the higher level propositions of external or transcendental law – on the other. Without this legal philosophy will remain only a non-committal speculation.

Besides, it should be mentioned that not only classical metaphysics as a general ontology but also ethics as a practical science must satisfy the formal stipulations of logic, the latter being, so to speak, the *conditio sine qua non* of each rationally supported argumentation.

Neither metaphysics nor ethics are empirically (i.e. through perception, or/and observation) founded. This explains the fact that these two basic disciplines of theoretical and practical philosophy have been since early antiquity compared to mathematics. Thus, Plato considered mathematics to be the propaedeutic or "pre-school" of metaphysics. Appropriately, the following words are said to be an engraved above the entrance of Plato's Academy: *Medeis ageometretos eisito* (those that are ignorant of geometry may not enter this place). The modern age has witnessed the ethics of Benedictus de Spinoza, written in the style of geometry (*ordine geometrico demonstrata*), presenting the fundamentals of moral behaviour in the form of definitions, axioms, and theorems brought together in an axiomatic-deductive system.

The ambition to create eternally valid and consistent system of ethical principles looks more promising compared to the systems of positive law that are constantly changing in history. Yet, due to stipulations with regard to its contents, this ambition has proven to be unfeasible. For the axiomatic-deductive method is based on the character of the axioms the validity of which is either evident, or not possible to prove, or requiring no proof, and which generate through deduction all further propositions as theorems.

Exactly the not beyond doubt evidence of the first principles has been responsible for the failure of all axiomatisation (with regard to the contents) attempts in ethics. What remains today as relevant to legal philosophy is the conception of the formal justification of the system, requiring logically consistent relations between the legal norms applied in the logical argumentation of a juridical decision. A weaker logic, however, could be a sufficient means to this end. The logic used nowadays with regard to ethics and philosophy of law is the so-called deontic logic, which demonstrates the consistent and non-contradictory relations between any normative propositions:

- The deontic logic refers to propositions expressing not what "is" but rather what "ought to be" (Greek *to deon* = the ought). Another name for it is logic of the rules of action (*Logik der Handlungsregeln*) or shortly – action logic (*Handlungslogik*).
- The deontic logic is a modal logic of ethics, e.g. it is based not on the concepts "true" and "false", but it rather refers to the concept of modal logic "possible", "impossible", "necessary" transferred on the level of human action as "permitted", "forbidden", "imperative".

Nevertheless, the deontic logic doesn't replace either the ontological or the axiological dimensions of law. That is, it doesn't say how one – on the basis of particular (content defined) norms and values – ought to behave. Rather it outlines the formal conditions leading one to a formally correct judgement about permitted, forbidden or imperative patterns of action.

Georg H. v. Wright has developed the deontic logic to avoid the relativity characteristic of all existing previous attempts to axiomatise ethics. The application of the notion of truth present in both common logic and in modal logic (for "possible" is "possibly true", "impossible" – "impossibly true", and "necessary" – "necessarily true"), leads in ethics to a theory of objective values. Thereby some presuppositions of logical method are to be actually explained only beyond logic and within meta-ethical research. Traditionally, this meta-ethical research has always been regarded as metaphysics or ontology, e.g. as theoretical philosophy coming before practical philosophy.

At least since Kant, however, in the sphere of ethical justification, the primacy of practical reason over the theoretical one has been maintained. Deontic logic provides a modern logical tool in the creation of ethical argumentation strategies without explicit ontological-metaphysical pre-decisions. In this sense, it also provides the foundation for a practical discourse theory, which offers a model of deeper justification of legal norms, reaching beyond the linguistic discourse theory without regressing to an objectivistic metaphysics or ontology.

0.1.2. *Ontological Dimension: Metaphysics and Law*

Nowadays one is almost always linking the concept of an ontological dimension of law with an "objectivistic metaphysics". This is illustrated by the words of one of the main representatives of modern ontology of law – Kaufmann. He claimed that ontology was a philosophy built upon the trust into the being. Thus, it presumes the existence of a being, which is not dependent on our thinking – e.g. objectivistic metaphysics. This philosophy devotes itself to the fact of being and not to consciousness. Man is not the measure of the things but vice versa (Kaufmann, 1984, p. 191).

> Such an ontological position allows for an objectivistic philosophy of law assuming that it is the being as such that carries in itself order and shape (*Gestalt*), that there exists a natural order of things and relations, that law has accompanied life in the human community since the very beginning (Kaufmann, 1984, p. 191).

Hence, the modern ontological efforts to justify law are explicitly against subjectivism, i.e. they attempt to establish law on an independent of the human beings basis. All these general formulations, however, have failed to name this basis.

Thus, the notion of ontology is burdened by historically different possible interpretations. The term "metaphysics" had been in use already at the turn of the 5th century (applied to Aristotle's *prima philosophia*). On the other hand, the notion of ontology gained popularity not earlier than the New age. Its only function was to represent the pure or general part of metaphysics as a study of being or the foundations of being. In accordance with the Aristotelian-scholastic tradition, the special part of metaphysics consists of

- Rational cosmology,
- Rational psychology, and
- Rational theology.

In the course of modern development of philosophy, ontology detached itself from rational theology, which at the time of Aristotle and the scholastics was inseparable from general metaphysics. Exactly that form of metaphysics – understood as *ontologia generalis* – was attacked by Kant who considered the being or the thing in itself for unrecognisable.

Compared to classical metaphysics, the alternative ontologies after Kant are weaker. Here belong:

- Hartmann's doctrine of strata, which as "inductive metaphysics" regresses at the end to a mere study of leftover legal problems, unsolved by empirical sciences;
- Phenomenology (main representative Husserl) where ontology is not a metaphysics but rather a study of essence (*Wesen*) or of the meaning of abstract notions;
- Existentialism perceiving ontology not as metaphysics but only as a study of categories, e.g. a study of the "pre-given in the mind essence of being" (*von dem im Denken vorkommenden Wesen des Seins*) (Jaspers). Or, for that matter, Heidegger, who limits ontology to a presentation of the state of human existence without being able to reach the very goal, namely, the "elaboration of the problem of being in general".

Legal philosophy has been connected, more or less successfully, with all these recent ontological concepts. Hartmann's ontology of strata is a concrete presentation of the comparatively autonomous mental being as "objective mind". According to Hartmann, law has to be regarded not only as a complex of norms, externally binding the humans, but as an objective spiritual power as well which together with other similar powers like religion, art, and science shapes the individual (See Engisch, 1971, p. 144).

In his famous talk *Apriorischen Grundlagen des bürgerlichen Rechts* (The a priori Fundamentals of Civil Law) Reinach attempted to set up "laws of essence" (*Wesensgesetze*) in the frames of an *a priori* legal theory. The latter is an example for the consistent binding of legal philosophy to phenomenology. Reinach's idea is that the legal entity provides eternal laws, which do not depend on our comprehension – exactly like the laws of mathematics. Unfortunately, as pointed out by Engisch, a more precise analysis of such eternal laws of essence proves them to be analytical and thus tautological judgements, or they are not evident. An example for a non-evident law of essence reads as follows: The fact that a claim lapses through an act of renunciation is founded in the essence of the claim as such and thereby is necessary and general.

The non-evident character of this law is demonstrated by the simple consideration that the claim of a treatment worthy of the human being naturally could not be lapsed through renunciation, i.e. there are claims that cannot be renounced.

The linking of legal philosophy with existential philosophy proves to be only a weaker version of classical legal metaphysics in the sense of natural law theory. According to Maihoffer natural law could no longer be derived from the "generally alleged essential nature of man and the generally alleged essential order among humans". Rather it is a law reflecting the human mode of existence in the world (Quoted after Engisch, 1971, p. 220).

Hence the ontological foundation of law within existential philosophy could lead to nothing more but ethically correct attitudes and sentiments described with notions like essentiality, resoluteness, conscience, guiltiness, existential communication etc., but not to principles and norms, prescribing the correct legal "ought". Engisch outlined the situation quite adequately by claiming that while research from the point of view of existential philosophy linked the "just" law with the concrete mode of human existence it nevertheless didn't determine it (Engisch, 1971, p. 221).

Or, the ontological foundation of law, pursuing the essence of law and not a particular modality of being – i.e. this of humans – would inevitably fail if it didn't refer directly to the "ontological structure of law" (Kaufmann). With the conception of ontological difference Heidegger, one of the founders of existential philosophy, claimed practically the same:

> The first philosophical step towards understanding the problem of being is that the being as such should not be determined by reducing it to another being (Heidegger, 1953, p. 6).

Traditional metaphysics accounts for this ontological difference through the distinction between "existence" and "essence", methodologically resolving it by the *analogia entis* doctrine. By claiming that all regional spheres of being show substantial similarity, *analogia entis* permits to make out the being of law by comparing it to the human mode of being. However, one has to maintain the distinction between essence and existence in order to prevent the transformation of classical ontology as *essentia* theory into a regional real ontology, e.g., into sociology or social anthropology. As Kaufmann, one of the main proponents of the ontological school in modern legal philosophy, stated, the establishment of the ontological structure of law was always a matter of "fundamental ontology" (Kaufmann, 1984, p. 116).

Hence, the permanent goal of legal ontology has always been to prove the objective validity of law before empirically experiencing its efficiency in human communities. The contemporary interpretation of the ontological dimension of law prevents the mutual of essence and existence, of essence and legal positivism. In this it differs from rationalistic metaphysics and *a priori* legal philosophy, which – also in contradiction to the traditional Aristotelian-scholastic *analogia entis* – tries to construct beyond the positively existing law a pure (e.g. independent of the empirical legal reality) theory of ideally valid law.

This means that in the historically changing positive law there exists an ontological 'hard kernel', which is not to be eliminated. Yet, this kernel could neither be derived from a general metaphysics of essence (ontology) nor could it be elaborated on the basis of a phenomenology of essence or on the basis of the mystique of existential philosophy of being.

The ontological dimension is not only a necessary subject-matter of the ontological school in contemporary philosophy (e.g., Kaufmann, or neo-scholastic legal philosophy etc.), it is also directly or indirectly present in all modern trends of legal philosophy as pinpointed by the commonly adopted notion "nature of the things". Methodologically speaking, the ontological dimension of law possesses an intrinsic function. Namely, it can be compared (within the framework of philosophy of science) to the indispensable for empirical knowledge function of the so-called "metaphysical background knowledge", the latter presenting a non-falsifiable hypothesis of a higher level. The difference here is that in the case of the ontological dimension of law the hypothesis relates to the "ought" and not to the "is". Hence, the legal ontological dimension reveals itself in a hypothesis of norms (Kriele) superposed on positive law, thus being always linked with the axiological dimension or ethics.

0.1.3. Axiological Dimension Ethics and Law

Serving methodological goals, legal positivism has separated law and morality, thus making the existence of an axiological dimension of law impossible – at least in the sense of traditional systematic philosophy where ontology and axiology are irrevocably connected. Still in Plato's doctrine of ideas, the idea of good is juxtaposed with the idea of truth. A direct reference to the axiological dimension of law is traced back to Aristotle, who further developed Plato's idea of justice by claiming that the biggest good possessed by a state was the "rightful", devoted to the existence of all. However, such an explanation of justice only reflects the formal justice as a relation or proportion, which later in the activity of the judge would reveal itself as a specific "adjusting" or "corrective" justice. This formal idea of justice, expressed by the notions of equality, proportionality and equivalency has been present in all stages of its own development and transformations throughout the whole history of legal philosophy – from the Middle Ages (Thomas Aquinas) to the New Age (Leibniz to the present. The most persistent in this idea proves to be the distinction between subjective and objective justice: Subjective justice is the confidence in the realization of objective justice, the former relating to the latter as truthfulness to truth. Hence, objective justice is the primary form of justice and subjective justice the secondary one (Quoted after Engisch, 1971, p. 155). Being quite abstract, the idea of formal justice has to be "filled" with value oriented contents in the process of its concretisation (Engisch). At this point, modern legal philosophy turns to the axiological dimension of law, making it its central problem thus restricting itself to legal ethics. Thus, a workshop on legal theory worked out a scheme relating theory of law to the logical structures of law, sociology of law to the sociological structures, and philosophy of law – to the axiological ones (Kaufmann, 1984, p. 10). Being at first glance clear and persuasive this systematisation could hardly be implemented in reality, e.g. the reduction of philosophy of law to a mere legal ethics is argumentatively grasping in thin air. In other words, a philosophy of law without logic and cognitive theory can be in no way considered a systematic discipline.

Assuming that the axiological dimension of legal philosophy could be dealt with only in connection with or under the stipulations of the methodological dimension, one can clearly differentiate between the notions of principles and values, which in juristic context are commonly treated as the same.

An example in this aspect is the *Spiegel*'s (an Austrian magazine) ruling of the Federal Constitutional Court 20, 162: Instead of declaring that "the freedom of press ...implies the possibility of developing a conflict with other values acknowledged by the Constitution", the Court could have ruled out that the freedom of press implies the possibility of colliding with other principles established in the Constitution. This alternative is not really a farfetched one as the further Court effectuations show: one the one hand, the Court proclaimed the "principle of freedom of press", on the other, however, it referred to the "certainly no less important constitutional principle of absolute duty of all state organs, institutions, and citizens to protect the stability and safeguarding of the state and its liberal order" (Quoted after Alexy, 1985, pp. 125-126).

Despite of this mutuality, based mainly on the fact that the gradual realization of values corresponds to the gradual accomplishment of principles (Alexy, 1985, pp. 125-126), it should be emphasized that both levels have to be kept methodologically apart:

- The principles are commands and as such they belong to the deontological sphere.
- By comparison, the values refer to the "good" and hence, they belong to the axiological sphere (See Wright, 1963, pp. 156-157).

In the history of ethics this distinction accounts for the classical controversy between formal ethics and the ethics of material values.

Main proponents of the ethics of material values are Max Scheler and Nicolai Hartmann. However, as an aspect of the discussion in legal philosophy, this ethic was first introduced in Welzel's *Naturrecht und Materiale Gerechtigkeit* (Natural Law and Material Justice).

Of interest to the contemporary discussion in legal philosophy is Hartmann's claim that while the first, non-deducible values – i.e. the autonomous basic values like "good", "personal freedom" etc. – are identical with the principles, the derived values, being for him also "purely ideal essences" (e.g. the broad sphere of mean values) are nevertheless no principles.

Scheler acknowledged Hartmann's analysis of ideal and normative ought as a valuable refinement of his own contemplations of "values and ought". The fatal view about the relatively "strong" ethical categories of lower level of hierarchy and the relatively "weak" ethical categories of higher level also developed in this context. Thus, Scheler claimed that "...the higher the values in the hierarchy are, the smaller is the possibility to realize them through will and deed" (Scheler, 1954, p. 12).

Additionally, one must differentiate between "principles" and "values" on the one hand, and "interests" and "needs" on the other. The latter, according to Wright, relate to the anthropological level and do not fit in the pure philosophy of law.

0.2. The Traditional Basic Controversy (Natural Law Versus Legal Positivism) and Contemporary Approaches to It

In keeping with present-day theories of law, it is the study of natural law on one hand and legal positivism on the other that define the traditional basic controversy of philosophy of law since its very beginning (See Schramm, 1978, p. 59).

No matter how irreconcilable, both viewpoints have nevertheless a common methodological starting position: The strict differentiation between "is" and "ought" is usually traced back to Kant (see Kaufmann, 1975), but it in fact represents the positivist thought. Thus, both, legal positivism - at least in the form presented in Kelsen's *Pure Theory of Law* – as well as the rationalist doctrine of natural law assume that norms' foundation is totally independent from the concrete historical reality. Both schools agree that the correct method to reach the right decision is through deduction from these "basic norms".

What makes the two traditional schools of philosophy of law irreconcilable alternatives should be searched for at another level: The theory of natural law assumes the existence of absolute and ever valid legal norms founded in "human nature". This concept of practically only ontologically (in the spirit of classical general metaphysics) argued idea of a transcendental natural law has always been contested by all known forms of legal positivism. Although accepting that legal constructions depend on historical changes, legal positivism denies the broader view of the historical character of law maintained by the representatives of the historical school ever since Savigny. Being "normativistic", modern legal positivism claims that all concrete legal decisions are deduced from norms, the latter containing the former.

Thereby – practically, if not also theoretically – legal acts and law are equally situated, at least in the sense that legal science doesn't need to deal with further reaching questions of the ethical foundation of legal positivism. Kelsen claimed this view on the basis of a comparison between normative and causal sciences. In the first instance this comparison leads to the demarcation between legal science and sociology. However, Kelsen's understanding of empirical science was based on natural sciences. Accordingly, he located sociology in the field of causal science where the sensually perceived phenomena were connected according to the law of cause and effect. Yet normative sciences – among them legal and moral sciences – do not organize their phenomena according to the principle of causality, but rather to the principle of imputation. Linguistically, this difference is expressed by the copula "ought" instead of "is" (or "will be").

This point of view has become neither obsolete nor irrelevant. Actually it expresses – as stated by Dreier in his remarks to Kelsen's Pure Theory of Law – "a still dominating matter-of-factness of continental European jurisprudence" (Dreier, 1981, p. 219). This has – one should add – a legitimate philosophical core based on the fact that the questions of ethics or moral sciences as normative sciences cannot be replaced by empirical research and explanation. This is not to say that legal science as such has nothing to do with sociology. Rather, being a normative discipline, legal science exercises methodological interest in its foundational structures on another level. Kelsen differentiated quite strictly between these two levels only to bring them together again (although there exist other interpretations of this aspect). Thus, his pure theory of law describes law as it is, and not as it should be according to some idealistic phantasms. However, the "is" of legal norms is compulsory, resulting from the "normative coercive order" of real human behaviour.

For Kelsen the difference between "is" and "ought" is not ontological, but a formal-logical one, as he himself claimed. Without leaving the borders of the formal-logical approach, one cannot find the way leading from one to another and both worlds remain separated by a gap not to be bridged. However, this doesn't mean that in legal practice (in the terms of the pure theory of law understood as a theory of positive law), the social effectiveness of a norm established by the authorities is of no interest for the evaluation of its legal validity. Rather, as Kelsen himself emphasized, a norm is legally valid if and only if it is *in toto* socially effective, i.e. if it is applied, followed and recognized.

This hint at social effectiveness brings us in any case to the practicability of norms beyond their validity in terms of formal logic. Yet, this formulation has proved to be too weak to eliminate the fundamental positivist thesis that formally "any content" could be law.

Unlike Radbruch, Kelsen continued to advocate that claim in his *Pure Theory of Law* (1960), even though at that time it had become long evident that any totalitarian regime led *ad absurdum* the view that law was only "what those who represent the state power postulate, according to their free and unlimited judgement" (Weinkauf, 1970, pp. 54-56). Kelsen's distinction between "is" and "ought" (as a formal-logical distinction), is to be found at a level different from the level of socially or politically desirable or acceptable law. The consequence of this view, however, is the still present belief in the vein of legal positivism that law and morality and, therefore, legal science and ethics are to be separated.

0.2.1. Legal Ethics Without Metaphysics

If one tries to understand the classical and still current thesis of the separation of morality from law – assigned to each form of legal positivism – one should fundamentally dismiss the idea of legal positivism as a homogeneous theory with homogeneous principles.

Its best known contemporary representative, Hart, distinguishes between five – at least – claims which are generally characterizing legal positivism, but which could be used only each for itself, being not recognized in their entirety by any of the advocates of legal positivism in the past or now:

1. Laws are human commands.
2. There is no necessary connection between law and morality, i.e. law as it is and law, as it ought to be.
3. The analysis (or the examination of the meaning) of legal concepts is (a) worthwhile, and (b) it is to be distinguished from historical research of the relation between law and other social phenomena and from the critical appraisal of law from the stand-point of morality and other social ideals.
4. Any legal system is a "closed logical system" allowing for correct legal decisions by means of logics, i.e. the deduction of legal decisions from pre-existing legal norms without reference to social goals, general rules, or moral standards.
5. Unlike propositions concerning facts, moral judgements are not to be justified by rational arguments, reasons or evidence: ethical "non-cognitivism" (See Hart 1971).

From a historical point of view, Bentham and Austin were the first to advocate claims 1 to 3, but they rejected the rest. However, the first three claims are doctrines, which are logically and with regard to the contents independent from one another. The fundamental assertion of the so-called imperative theory of lawconsists of commands of the political sovereign. But this is neither the logical consequence of nor the prerequisite for the demarcation between law and morality. If they have to be separated at a conceptual level, then the question about the basis of validity of positive law remains open. What is clear is merely the fact that law is somehow connected to the notions of power, coercion or violence.

In his fundamentally modified imperative theory, Kelsen pointed out that it would not be correct to characterize norm in general, and the legal norm, in particular, as will or command of the legislator or of the state if "will" and "command" were looked upon as psychological acts. That is, the fact that somebody is giving a command is not a reason enough to consider this command as valid, i. e. as coercive for the addressee norm (See Hoerster, 1977, p. 25).

Thereby, one is confronted with the problem of the competence, which should be possessed by the authority establishing the norms. According to Kelsen such a competence can be based only on a norm, giving the prerogative to establish norms. Or, to put it differently, there should be at least one norm which itself is not a command of the authority. The empowered to establish norms authority is subjugated to this one norm as well as the norms requiring obedience towards the norms, established by the respective authority. This norm, subordinating all other norms functions as a higher norm, as the "last highest" norm in fact. It has to be always presupposed as far as no authority could establish it. This one norm is called by Kelsen basic norm: The basic norm constitutes the unity of a diversity of norms, being the reason for the validity of all norms belonging to the system (Hoerster, 1977, p. 26).

Therefore, law is a coercive system with an optional content. The difference between law and other coercive systems, i.e., the one of a hold-up gang, is to be looked for not in its contents, but rather in the existence of such a basic normHold-up gangs, or the pirate states using to bring insecurity to the North-West African coast in the 16th - 19th centuries, do not have as a premise this basic norm. Confronted with legal systems possessing a basic norm and because of that being more effective, they disappeared by gradually loosing their members who were either brought to prison or executed.

Consequently, Kelsen's formal, and – with regard to the contents – "empty" legal theory could hardly answer questions about the contents of the basic norm of a state legal system. Thus, Kelsen claimed that coercion from man to man should be exercised in the way and under the conditions determined by the historically first constitution. The basic norm delegates to the historically first constitution to determine the procedure setting the norms, which establish the coercive acts (Hoerster, 1977, pp. 33-34). However, if one doesn't accept Kelsen's legitimisation on the basis of a merely hypothetically pre-existent norm, then the systematic question has to be transformed into a historical one about the origin of the first constitution.

A step even further away from the original imperative theory of law is the view of Hart, who perceives the norms as socially acceptable rules, bindingly recognized by the citizens for regulation of their mutual existence.

This view corresponds to the 5th claim – the non-cognitivism, – asserting that moral judgements could not be rationally justified.

Legal positivism has not yet answered the question how far, or better to say, how deep the rational argumentation of fundamental norms or principles, respectively, values should go. Unanimous, however, is the opinion that such an inter-subjective ethics has to be realized without metaphysics (Hoerster, 1982, pp. 246-272). Still, this isn't a total rejection of the options of metaphysic. Rather, this opinion reflects the assertion "that there exist no normative propositions ("ought"-propositions) which can claim to have the status of a synthetic knowledge *a priori*, and therefore objectivity".

However, this would eliminate the ontological dimension of legal philosophy. The so-called hermeneutical legal ontology makes its concern to maintain this dimension without regressing into a naive objectivism.

0.2.2. Through Natural Law and Legal Positivism Toward a Hermeneutical Legal Ontology

Although legal positivism (in the general sense of the word – as referring to the positively staid law) has existed all along, it is nevertheless the natural law theory to appear as historically the first systematized doctrine of law.

Thus, contemporary natural law theory owes its persuasiveness to the traditional systematisations of the mediaeval scholastics, culminating in the *Ordo* concepts of the Thomistic ontology. In this sense, one could say that the natural law theory of our century is bound to a renaissance of the Aristotelian-scholastic philosophy. Neo-scholastic and neo-Thomism mark the frames of a "Christian legal philosophy". As a matter of course, compared to theology the latter shows confessional differences. Thus, the catholic natural law theory (Messer 1950) co-exists with an evangelic theological ethics and legal philosophy (Thieleke 1958, Brunner 1943). Common for all these conceptions is that legal philosophy (as legal ontology and legal ethics) necessarily flows into the sphere of legal theology. As Wolf stated: It has to, because both legal ontology and legal ethics could neither raise nor answer the question about the "where from" and "where to" of the legal "is" and the legal "ought"... the legal "is" exists in humans through the affirmation of God, humans can lawfully exist only before God (Wolf 1972, p. 83).

Methodologically this is also reflected in the model of argumentation of Fikentscher. Referring to Kaufmann's words: "Going two steps deeper in the argumentation will transform each legal question into a religious one", he himself claimed: "Everyone who judges needs a *religio*, an Archimed point to support the lever of his judgements" (Fikentscher, 1975-77, p. 405). Voices against such a direct binding of legal philosophy to a particular religion, a religious world view, or a particular theology have been justifiably raised. It is true that each religious attitude carries in itself judgements. The opposite, however, namely, that there is no judgement without a religion, is not necessarily valid. Here, at least, there exists no visible logical connection. What is more, there arises the problem of the "competing religious messages" leading to different systems of values (Hoerster 1982).

The ontological method of argumentation is not only "one venerable, classical method". It offers as well a variety of modern opportunities, leading from the New Age statements of law of reason to the reflections of phenomenology and existential philosophy to the empirical concepts so that the ontological position dissolves itself into an empirical and cultural-scientific one. The same could be said about the largely discussed notion "nature of the things". While representing the most popular form of modern ontological thinking in natural law, it is nevertheless, (or because of this), very much ambiguous. The notion has its roots in ancient natural law and is often traced back to Aristotle ("the determining mean") or to Thomas Aquinas ("the just thing itself") (Kaufmann, 1983, p. 173). Already in these formulations it possesses a concrete character of a transcendental natural law as *lex aeterna*, of a "just law". Or of an abstract "natural law of general legal maxims" deduced from human nature. Because of the latter it is also known as a "concrete natural law" (Maihofer, 1958, p. 146).

The above developments have been provoked by Radbruch's contribution to the commemorative volume in honour of R. Laun in 1948 (p. 283-285.). According to Kaufmann (1984, p. 173), Radbruch paved the way for a "new legal ontology". Also that put an end to the era of legal positivism in the strict sense of the word.

What could be said about this new legal ontology, which – to follow the unanimous opinion of its representatives (Maihofer, 1962, X) – should become the new ontological foundation of law after natural law and legal positivism?

The question is to be answered from two perspectives – with regard to the contents, and methodologically:

- It can be no longer based on the controversy between strict wording of law on the one hand, and legal norms existing beyond law, on the other.
- Methodologically spoken, it can be no longer based on an epistemologically naive view of ontology.

Being quite general this negative answers permit a variety of more concrete solutions, which could be linked to philosophically different basic attitudes:

- The nature of things understood merely as a theoretical construct (*Denkform*), in the sense of Radbruch.
- "Structures, logically adequate to things" in the sense of Welzel and Stratenwerth.
- Endowed by God, pre-given institutions in the sense of Forsthoff's "institutional legal doctrine".
- "Ontological structures" of an action and the resulting social roles as, e.g. shown in Luhmann's system-theoretical legal sociology

Still, all these attempts to answer the question about the new ontology presume the existence of "pre-formed realities of values" (Kaufmann), which determine and restrict legislators' as well as judges' decisions with regard to the contents. Also they represent the basis for a later critique of these decisions. In addition, this particular aspect demonstrates the necessity of legal metaphysics: To do without it would mean to leave the idea of just law to the decisions of political powers. Thus, the resistance legislation (inevitably rejected by legal positivism in its older as well as newer coinage) has been mainly supported in such trans-legal instances like natural law, individual conscience or religious convictions (Kriele, 1979, p. 11).

According to Kaufmann (1984, p. 81), what makes the new legal ontology different is that nowadays the "rightness" of law should not be taken for granted. Rather, it should be seen as problematic and open to revision. The idea that there is no unquestionable "just" law dissociates the new legal ontology from natural law theory and legal positivism, which – each in its own way – assert the validity of legal norms as apodictic thus forbidding critique external to the system.

The adequate method for this modern legal ontology is the one of hermeneutics. It argues the rightness of law not on the basis of pre-existing, trans-legal and ever valid legal norms, but rather on the basis of the dialectic between facts of life and norms, between "is" and "ought".

In this way, modern legal ontology (as hermeneutical legal ontology) pursues to avoid the reproach of objectivism. Objectivity in the sense of absence of subject could not be considered a possible ideal of jurisprudence. The judge deals with norms, which cannot be recognized and ordered in a subject-free "purity". Rather, they are connections and relations, which have to be established always anew. Modern hermeneutical legal ontology refers to ontology only (if at all) in the sense of a "relation-ontology" and not in the sense of classical "substance-ontology" in the Aristotelian-scholastic tradition.

However, it is difficult to distinguish this weaker form of ontology from the mere functionalism of the entity of each being's relations. Such pure functionalism is advocated in the system-theoretical legal sociology (Luhmann), which rejects the question about just law as a superfluous one. The legitimisation of law is realized through the proceedings themselves. What differentiates this legal sociology from hermeneutical legal ontology (claiming that no just law could exist outside the decision-making process) is the essential re-formulation of the problem. According to it the rightness of law is ascertained in the procedures but it is not identical with the procedures themselves. What is, then, the type of "is"-reality (which constructs identity and which cannot be freely manipulated), referred to by hermeneutical legal ontology?

According to Kaufmann what we are searching for is the person as an intersection between object and connection, between static and dynamics. The relationship, the structural unity of *relatio* and *relata* is concretely realized in the person. The above mentioned, relation-ontology should be thus formulated as ontology of the personal, recognizing not the thing but rather the person as an example of "is". Hence, the ontology of law has to be constituted as a relation-personal ontology (Kaufmann, 1984, p. 98). Thus, what matters is not any longer the search for the "law in itself", but rather the particular personal links between people and things, and between the people themselves. Regarded as performed or omitted actions, these links become legally relevant. However, the relevance is not automatically presumed – first it has to be determined in each particular case. Thereby, the hermeneutical method acquires a transcendental meaning – it is perceived not only descriptively and psychologically as a mere understanding of the text but as having a constitutional character as well: A pre-understanding accompanies as an essential fundament the interpretation of law. This pre-understanding is hypothetical e.g. it can be at any time revised. Nevertheless, it is a conception of just law, which transcends positive law.

Emulating Kant, Joachim Hruschka (1972, pp. 10-12) named this conception "hermeneutical trans-positiveness of positive law". This reveals one more link of modern legal philosophy to the traditional concepts – the reference to Kant's transcendental philosophy is a reference to law of reason of modern times.

0.2.3. *From Law of Reason to Positivism and Back Again*

The third position in the vein of the traditional and basic controversy in legal philosophy is what Dreier determines as "from law of reason to positivism and back again".

Actually, it is the story of the reception of Kant's practical philosophy. According to Dreier, this reception – regarding the philosophically relevant aspect of law – has undergone three phases:

- In the first phase Kant's practical philosophy led to a series of compendia on natural law, showing the influence of Kantian ideas, e.g. T. Schmalz (1795) *Das reine Naturrecht* (Pure Natural Law). Königsberg. New Edition; Aalen, 1969. However, already at the beginning of the 19th century, positive law started to replace natural law; this became apparent in the so-called "jurisprudence of concepts". Nevertheless, natural law, respectively the law of reason of modern times remains as an immanent basis of positive law.

- In the second phase, the phase of Neo-Kantian legal philosophy, this conception was pushed aside to make place to a formalistic legal positivism (in Neo-Kantian tradition), strongly connected with the names of Kelsen and Stammler.
- The third phase is marked by a rebirth of the Kantian law of reason as a consequence of the natural law debate taken place after 1945. This rebirth reflects the coming together of the classical ontological view in natural law theory with the modern law of reason, best recognizable in the concept of nature of the things, revived anew by Radbruch.

The relevance of Kant's work for the contemporary discussion is to be found in the fact that his interpretation of the complex interrelations between nature, reason, and history overcomes (for his time) the classical controversy "natural law – positive law". Thus, according to Kant, reason is a practical instance, which should be understood as a historical phenomenon. Reason creates laws, which have to be seen not theologically or empirical-naturalistically as inborn, but rather as the result of free legislation, functioning in real history. If, then, Kant's philosophical doctrine of law is to be ordered in the tradition of natural law, one must consider nature (according to Kaulbach) as an idea in philosophy of history, helping reason to produce a picture of the options and perspectives in the realization of this reason's legislation. Therefore, Kant's philosophical doctrine of law is not – as also Dreier in argument with Kaulbach emphasised – a-historical "zero-theory" which claims that the realisation of the natural law principles has to start always anew. Rather, the reality these principles apply to, must be interpreted as a historical process. Or, as Dreier explicitly states: "A process of development, permitting progressive formation of institutions". Kant himself put the foundation of such a theory of historicity of law, regarding history of humans as the realization of a hidden plan of nature to create a state constitution as the only way to guarantee the development of all dispositions of mankind (Kant, 1986).

According to Dreier, this strong binding between practical philosophy and philosophy of history is what makes Kant's legal philosophy currently relevant. Hence, the "farewell" to Kant and Hegel proclaimed by Klug (Baumann, 1968, pp. 36-38) was somewhat premature.

0.3. Summary and Preview

To tackle legal philosophy as a systematic discipline would mean in the first instance to differentiate between its various dimensions and levels of argumentation, which are not of the same value. That is why, each systematisation of legal philosophy has to overtake their evaluation, ordering and integrating them at the end.

Following a phase of a "hopeless splinter" (Dreier) of legal philosophy, this integration becomes clearer on the background of a kind of "rehabilitation of practical reason". Thereby, legal philosophy reveals itself to be not a non-binding theoretical speculation, rather, it can be justified on the basis of practical grounds which transcend the creation of law by the legislator. Whether such a "theory of practical reason" is called legal philosophy or legal theory is of secondary importance and a matter of the nomenclature.

It is for sure, however, that the original restriction of legal theory to a pure theory of positive law can be no longer maintained. It has been also shown that a theory of positive law does possess its philosophical premises and implications.

The summary of the reviewed contemporary attempts to create a new systematic legal philosophy reveals several basic theses, presented below.

0.3.1. *Legal Philosophy With a Maximum of Methodological Reflection Within the Frames of Theoretical Reason*

The appearance of a legal theory, explicitly oriented towards methodology, and, consequently, of analytical legal philosophy and legal logic makes the traditional (particularly, Kant-oriented) legal theoreticians accept that a relapse of legal philosophy into a "pre-Kelsen's methodological syncretism" (Dreier), is no longer possible.

If nowadays, legal philosophy is to be treated as a systematic discipline, then only under the premise of a practice-oriented, adequate methodology. This would also mean that logical concept-analyses and logical reconstructions of juridical argumentation structures could not be the goal in itself, but rather the means to the solution of a concrete legal case. Classical formal logic usually suffices this purpose. In the cases, however, when the required logical argumentation transgresses the obligatory argumentation procedure, one needs further reaching formalizations in the sense of more rigorous calculations of mathematical or symbolic logic. For example, such are those instances when the validity of a conclusion is not simply taken for obvious, but has to be proven step by step. In this way, some typical for the juridical methodology types of reasoning as e.g., conclusion per analogy, or *argumentum ad absurdum* can be formally justified as valid ruling procedures.

Yet, at all levels of specification formal logic remains an analytical instrument, the efficiency of which can be pragmatically revised.

Accordingly, jurists often refer to Frege's comparison of formal logic to a microscope. Thus, on the one hand, the microscope makes visible what the naked eye doesn't detect, on the other hand however it is inferior to the naked eye in absorbing the variety of tasks.

Therefore, legal philosophy as well as juristic methodology (in the narrower sense of the word), cannot choose between classical Aristotelian, in natural language formulated logic on the one hand, and modern, symbolic or formalized logic, on the other. What counts here is more the old Aristotle's principle, that an educated person pursues accuracy only as long as required by the purpose itself.

What's more, the problems of legal philosophy as well as the practical juristic problems are not caused by the lack of formal specification in the closure procedure. Rather, they are due to the fact, that the construction of a simple by itself deductive inference is seldom possible when one takes an abstract clause out of the statuary text turning it then directly and without problems into a concrete clause, acting as a descriptive statement of a particular case. The true problematic nature of acquisition of law is to be found in one repetitive and ever concretising approach, viewing the goal through a "perspective, constantly changing" (Engisch) between the state of facts and the norm. This approach can be methodologically presented through a cyclic or spiral mode.

Yet, the logical analysis of legal concepts (*nomen juris*), which paves the way to the logical subsumption closure, proves to be of greater importance in the argumentation of a legal decision than the formal-logical specification of the closure procedure. Far from being always a pure classification, the legal concepts do not permit to declare strictly whether they subsume a particular case or not. Their complicated logical structure thus confronts the subsumption questions with the most difficult and by far the most philosophically loaded problem of finding of law. It is namely the basic norms of the highest level of a legal system which, being kept necessarily "open", require improvement and interpretation.

0.3.2. *Legal Philosophy as a Developmental Theory of Law With a Minimum of Metaphysics Within the Boundaries of Historical Reason*

Despite all restrictions undergone by legal philosophy as legal metaphysics since Kant, it has nevertheless preserved the task to "demonstrate the measure for the lawfulness of a positively valid law" (Kaulbach, 1979, p. 157).

The traditional measure for trans-legal law has been the so-called natural law or law of reason. Yet, the historical legal school (which with the reformulation of classical and medieval natural law into Kant's law of reason made a step forward against the static concept of law) has shown that nothing lasts forever in the human world. However, this school remained "the child of Enlightenment" (Kaufmann).

The recognition of the historical essence of the human "is" in our century has supplied the necessary arguments for the historicity of law, which – according to the rationalistic view – is rooted in human essence. This recognition has been accomplished not only in the sphere of natural sciences – e.g. through the broad acceptance of the evolutionary theory, but also in philosophy although in a completely different manner. Phenomenology and existential philosophy interpret the historicity of humans as the essence of the human mode of existence, subjected to history not only as a lapse of time. Rather, the "inner sense of this essence is determined through the fact that it has a history".

This is the distinction between the human being on the one hand, and animated and non-animated nature, on the other, the latter possessing no awareness of its history. Thus, only the human being has a history. Even the "absolute mind", existing out of time and history, and responsible for the divine natural law, knows no history of itself. A realistic reduction of the classical, religiously founded natural law to a law of reason will inevitably imply historicity of law through the historicity of human reason.

The modern proponents of the ontological branch of legal philosophy have been forced to recognize this fact, clearly demonstrated by phrases like: "natural law with alternating contents" (Stammler), "historical natural law" as "existential law of the essence" (Max Müller), "natural law with becoming contents to become" (*Naturrecht mit werdendem Inhalt)* (Fechner), or "natural law as historically adequate law" (Kaufmann) (see Kaufmann, 1984, p. 27).

Hence, at the end, the ontological dimension of legal philosophy transforms itself into a developmental theory of law. Its minimum of metaphysics should be the proof of reason in history, taken for granted that it will not be lost in historical relativism.

0.3.3. *Legal Philosophy, Sufficiently Founded on the Basis of Ethical Evaluation Principles Within the Frames of Practical* Reason

The thesis that norms could be and should be rationally founded, accepted even by legal positivism, forms the core of each modern legal philosophy, no matter how far the latter has removed itself from any ontological-metaphysical foundation. The model of sufficient foundation has originated from A. Pecznik's theory of non-deductive inferences, which explicitly allows for 'leaps' and transformations between different levels: In this particular case – between the level of juridical justification of legal decisions and the one of extra-juridical justification. That is, certain arguments in a legal context can satisfy a jurist, being at the same time completely unsatisfactory for a politician, or a moral philosopher. The sufficient juridical justification in the legal context is based on the existing juridical tradition, requiring in its turn a deeper reaching justification.

However, there is still no unanimity regarding the contents and the logical structures of this deeper justification, the possible models in this aspects being quite different. At the most one agrees that this process of rational argumentation can never be completed. Rather, it is to be interrupted again and again in the interests of the currently necessary decision.

From the point of view of philosophy of science, the models of deeper foundation reflect the concept of the regulative idea of truth: An absolute truth is not to be reached. However this is also not necessary. Up till now, there exists no scientific theory claiming the possession of the absolute truth; there exists also no legal system to claim the absolute and valid for all times law. What remains is the regulative idea of justice and then – only as a goal and never as a fact.

1. LEGAL PHILOSOPHY AS METHODOLOGY

Juridical methodology has a long tradition, going back to ancient times. Leibniz praised the classical Roman jurisprudence because of its precision (*subtilitas*) and compared it to the method of geometry.[2] This non-problematic methodological conception of jurists oriented towards classical logic was upset however in the 19th century when jurisprudence developed as a systematized science of law in accordance with the methodological ideal of modern (natural) science. Like Leibniz, Savigny attributed to the juridical procedure a degree of certainty "that can be found only in mathematics". (Savigny, 1840, p. 19) This approach was criticised by the persecutor Kirchmann in 1848 in his famous talk on the worthlessness of jurisprudence as a science. Kirchmann claimed that jurisprudence was not a science at all. In the meantime the literature concerning juridical methodology has become in fact formidable. Yet the question about the "scientific character" of jurisprudence is still of immediate interest.

This can be seen in the vast number of publications dedicated to the topic (See Wolf, 1953, pp. 27-29, Bockelmann, 1971, p. 29) as well as in the fact that concrete juridical controversies include many hints at the elementary rules of scientific knowledge (See Raisch, 1977, p. 29).

These rules and operations in science are central issues in philosophy of science, a discipline, which grew from the methodological reflections in all branches of scientific research. Philosophy of science systematizes these reflections and gives the particular methodologies of the sciences a general and broad "framework". Undoubtedly, the basic question of juridical methodology, too, cannot be "internally" solved and separated from the general problems of philosophy of science. Today's jurisprudence with its many specialized disciplines is to be found in a close contact with various fields of contemporary science. It is through reference to general philosophy of science that juridical methodology is linked with the methods used in other sciences; at the same time, the peculiarity of the concept of science in legal sciences can thus be explicated.

The idea that juridical methodology can be incorporated into general philosophy of science is not new. Already Kelsen referred to "general philosophy of science" maintaining this to be the way for the "methodological acceleration" of legal science (Kelsen, 1934). Likewise Larenz explicitly referred to "modern hermeneutics". However, this program has not yet succeeded, simply because, for now, there does not exist a generally accepted philosophy of science which, in its specifications, could

---

[2] Letter to H. E. KESTNER, July 1, 1716, in *LEIBNITTI opera omnia*, vol 4, p. 269.

supply each branch of knowledge with the adequate methodology. Therefore, irrespective of its reference to a general conception of what one calls "scientific knowledge", juristic methodology still includes two extreme positions. One is "scientism", a notion of science based exclusively on formal logic, mathematics, and empirical facts. The other is "hermeneutics" coming from the traditional notion of "culture" (*Bildung*) in the humanities. It has been always accepted that juridical methodology should necessary refer to the "true concept" of science (e.g. Kirchmann in the already mentioned speech). However this would require a general philosophy of science doing justice to the very pluralism of methods.

1.1. *Jurisprudential Concept of Science and the Position of Jurisprudence in the System of Sciences*

Introducing his well-known textbook *Methodenlehre der Rechtswissenschaften*, (Methodology of Legal Sciences) Larenz posed the following questions:

> What is legal science? Is it – as far as it is understood neither as historical, nor as social science, but rather as an autonomous discipline "jurisprudence" – science at all? Is it directed towards gaining knowledge, an intentional mental activity, or just an ordered *corpus* of knowledge accepted by a particular legal community *hic et nunc*? Or is it, perhaps, a kind of "technical arts", an instruction showing how to handle legal cases according to particular rules that are to be characterized as practical maxims or conventional norms allowing for even decisions? Or is it, perhaps, all this at one and the same time? (Larenz 1979)

In what follows I'll present the most important answers to these questions as far as they appear in contemporary discussions of juridical methodology.

1.1.1. *Legal Science as a Discipline of the Humanities or Cultural Sciences*

Ever since Dilthey and the neo-Kantians who – in German tradition – have separated humanities from the (natural) sciences, legal sciences have usually been subsumed under the former, also characterized as "cultural sciences" (Rickert). According to Engisch, this can be justified so: Legal sciences deal with "norms". Norms, however, are "constructions of the human mind". Therefore, legal science belongs to the humanities.

1.1.2. *Legal Science as a Pure Normative Science*

The extreme position of perceiving legal science as a pure science of norms leads to Kelsen's *Reine Rechtslehre* (Pure Theory of Law). In the preface to its first edition, Kelsen emphasized that from the beginning on it had been his goal to bring legal science "to the level of a true science, a

Geisteswissenschaft." In fact, unlike Engisch, he perceived *Geisteswissenschaft* as a formal and structural science thus using a "scientistic" notion of the humanities (See Larenz, 1979, p. 87). Different from this position is Klug's formal juristic logic which analyses (by means of modern formalized logic) the specific types of juridical inference, like analogy and reverse conclusion, *argumentum ad absurdum*, etc., and does not claim to round up the whole compass of juridical methodology. Formal and logical studies of juridical argumentation and proving, then, are independent of any theoretical and philosophical position regarding law.

1.1.3. *Legal Science as an Empirical Nomothetic Science*

Although in the 20th century no jurist or legal philosopher has placed legal science in the domain of the (natural) sciences, there have been attempts to separate legal sciences from the humanities and to bring them closer to the empirical sciences. It has been argued that in the sphere of the humanities the "individual" has the upper hand and no general laws are established, while legal science is "like natural science a science of laws" (See Engisch, 1977, p. 9). These laws are indeed not to be explicated like natural laws through induction and generalization. However, they should always be in accordance with the given social reality – its single cases must be subsumed under the legal norms.

Thus, in a way, legal norms have a hypothetical character and are not absolutely valid. The conditions determining the validity of legal norms are stated on the basis of certain empirical facts. Each legal statement is therefore an empirically founded norm. It relies on the social reality of law, which always consists of two parts:

- The empirical fact, and
- The legal consequence regulated by the norm itself.

1.1.4. *Legal Science as a Social Science*

Since the empirical part of the legal norms refers to a social reality, the legal sciences have usually been subsumed under the comprehensive system of "social sciences". In its extreme case, the idea of legal science as a special type of social science leads to the reduction of the whole corpus of legal sciences to sociology of law.

An early example is Eugen Ehrlich's *Grundlegung der Soziologie des Rechts* (Foundation of Sociology of Law) published in 1913. In this book the legal regime is considered as historically and socially given, and the legal clause, the norm as derivable from the "facts of legal life".

1.1.5. *Legal Science as a Practical Science of Action ("Jurisprudence")*

The characterization of legal science as a branch of the humanities or a particular kind of empirical cultural or social science does not really grasp its peculiarity and its difference from all the other sciences that refer only to goals of knowledge. Engisch determined this specificity of legal science as follows:

> It is the almost unique advantage of legal science ... not simply to go with or behind law, but to participate in the formation of law itself and of life in and under legality (Engisch, 1977, p. 8).

According to this view legal science is a practical science that does not only refer to knowledge of law, but also to its application and further development. Thus, there is no difference between legal science in the narrower sense of description, systematisation and interpretation of law, and "jurisprudence" as a system of juridical action and decision-making.

1.1.6. *Methodological Consequences*

In the contemporary methodological debate the classification of legal sciences as a part of the humanities, as a pure structural science, as a social science or as a practical science of action and decision is connected with a particular program in philosophy of science:

- The characterization of legal science as a branch of the humanities means a reference to hermeneutics as a method of "understanding" (*Verstehen*).
- The specification of legal science as a pure formal, structural science appeals to the scientistic-analytical logic of science as a method of deductive "explanation".
- Finally, characterizing legal science as a practical science means a reference to methods of action and decision, as they have been recently developed in the so-called theories of action in the context of general system-theoretical reflections.

Significantly, in today's juridical methodology none of these positions is advocated without any objection. The connection of the traditional hermeneutical stance with action-theoretical and decision-theoretical operations has already been established by the topical problem reflection. Also, the contrast between "scientism" (analytical philosophy) and "hermeneutics" is nowadays regarded as a mere "controversy of academic schools". Rottleuthner put it as follows:

The analysts have criticized the advocates of hermeneutics (and in fact not only them) by saying that they talk in a blurred manner and confuse everything. At the same time, the analytical hares range over fields of problems, and the hermeneutical hedgehogs are waiting at the balk and tell the analysts that they have long before said what they (the analysts) elaborated only with some difficulty; and that they have learned something about terminology, but nothing that really matters. The game is reversible: The analysts, this time the hedgehogs, are now expecting the hermeneutical hare to tell him that he was running through the field in a wild and confused manner. – It seems that in the controversy between two academic schools the opponents often want to play the role of the hare. But one should keep in mind that the hedgehog couple has already operated scientifically (Koch, 1976, p. 28-29).

In the following sections I want to show that the extremely elaborated methods of rational knowing and acting in the different sciences – e. g. in the humanities' method of understanding (*Verstehen*) or the method of explaining (*Erklären*) in the empirical (natural) sciences – are to be brought back to a common basis which has always been present in the pre-scientific common-sense knowledge. On principle, scientific knowledge is not different from common-sense knowledge, but it represents a refinement and systematisation of the latter.

1.2. *The Basic Types of Scientific Method*

If scientific knowledge is nothing else but a "sophistication" of common-sense knowledge, then the methods of science, too, should be pre-formed as "archetypes" at the level of everyday knowledge. Otherwise, one would not be able to explain the origin and the constant expansion of science. Each true innovation comes indeed from a sphere outside the respective scientific system, and is usually formulated in ordinary language before it paves its way into a scientific discipline by means of the constructive extension of the particular scientific terminology.

The starting point of philosophy of science is the analysis of the fundamental knowledge operations that constitute science:

- Description of objects and facts,
- Explanation of events, and
- Understanding of action.

With regard to their basic meaning these operations appear in all complex knowledge processes; however, they gain their accuracy and specific relevance only in the system of sciences.

1.2.1. *Description*

Description is to be considered as the first and elementary phase in any objectified or communicable knowledge process. It is based merely on the elementary knowledge operations, perception and observation, and thus creates the empirical material for both explanation and understanding.

- The simplest everyday type of description is the description of an object. One describes the properties and characteristics of the object in question.
- Since any object is always connected with other objects thus constituting facts, one usually needs a more sophisticated description, which is not just directed to single objects, but also towards their "relations".

By his doctrine of categories Aristotle – long before modern science – established a model of testing and verifying the completeness of description. Categories are to be understood as possible statements that can be basically made about objects and facts. Aristotle formulated them as questions, so that they possibly came from the sphere of juridical reconstruction of facts, i e. the judge's questions to the eye-witness. They supply a complete catalogue of the description of objects and facts.

Starting with the question 'what is it?' (e.g., a man, a horse), we have to go on in the following manner: How many (how much)? (e.g., two, three inches), how in relation to? (double, half of it, bigger), where? (e.g., on the market), when? (e.g., yesterday), position (e.g., lying, sitting), having (e.g., shoes, weapons), action (e.g., cuts, burns), suffering (e.g., is cut, is burned) (Aristotles, Organon).

This catalogue does not only serve as a model of description of objects and facts, but – as the question about time suggests – also as a model of describing events. The last two categories extend this model to the description of action.

Hence, Aristotle's doctrine of categories reveals itself as the classical fundament of a theory of description that has ever since remained basically unchanged in the further development of the sciences. Both, modern sciences as "descriptive natural science" or natural history, and the descriptive theories of action in social science and the humanities (particularly history), are based on this model, which then gives a methodological ideal.

Also, modern analytical theory of action is implicitly included in the Aristotelian doctrine of categories. According to Aristotle's model the descriptive elements of an action are the following:

- Subject of action ("who has done it?").
- Type of action ("what has he/she done?").

- Modality ("how has he/she done it?")
  (a) Manner and way ("which way has he/she done it?"),
  (b) Means ("by what means has he/she done it?").
- Context of action ("what's the context of his/her doing?")
  (a) Aspect of time ("when has he/she done it?"),
  (b) Aspect of space ("where has he/she done it?"),
  (c) Circumstances ("under which circumstances has he/she done it?") (See Rescher, 1977, p. 1-2).

However, a clear distinction must be made between the descriptions of the action itself and the question "why?" ("Why has this been done?"). The same is true to the description of events. An event is descriptive only as a change of facts, i.e. an alteration of the state of objects that are particularly related one to another. Therefore, the event is represented only by the description of single states and their sequence. An event can happen or can be caused by an action.

If one is ask to "What are the causes of an event?" the answer is an *explanation.* To ask "What are the motives of an action?" would mean to try to *understand* the action. With regard to the juridical factual statement – its context to the Aristotelian doctrine of categories is historically evident – the basic rules of the complete description are still relevant.

1.2.2. *Explanation*

In everyday parlance the term "explanation" is ambiguous. In its widest sense it refers to different statements and cannot be clearly separated from the notion of understanding. Each question regarding understanding can be formulated in such a way that the answer, after all, is an "explanation".[3] An example is moral reproach and justification: "I don't understand, why have you done this? Can you explain?" (Hübner, Menne, 1973, pp. 21-23).

Nevertheless, the very meaning of the notion of (scientific) explanation has been pre-formed in everyday language. In our daily life unpredictable events require particular explanations. Let us think of sickness and disease (contrasting health) or traffic accidents and catastrophes. But the same is true to scientific *causal* research:

- Astronomy emerged from explanations of irregularities of planetary motion and lunar and solar eclipse.

[3] The German word "erklären" or "Erklärung" has different meanings from the English explain or explanation. On this problem see Patzig (1979) Erklärung. In: Wissenschafts-theoretisches Lexikon. Cologne-Graz.

- Physics is rooted in mechanics, the "wonderful" lever effect – "a small force is moving heavy objects" (See Oeser, 1979, p. 14-15).
- Evolutionary theory is not based on the unchangeable, but on the modification of species (See Oeser, 1974, p. 79-81).

In science and in everyday knowledge the basic structure of explanation is nothing else but the reduction of the unknown to the known. Everyday knowledge considers an event explained when a known object, fact or action could be given as its natural or concrete cause.

Scientific explanation is different – but only because of its generality characterized by the statement of a general law that has its validity not only in the case of a single event but can be applied to the type of events. This way, an incidental event appears to be a necessary or probable one and the unpredictable becomes predictable. In scientific knowledge this means that each true, i.e. *correct*, explanation is – and has to be – potentially a prognosis. A decisive moment of any correct explanation is the precise description of the event in question. The description offers the material basis to explanations. If the material is not taken into account, we get only a pseudo-explanation. Examples are, among many others, the bad horoscope of a car driver who gets involved in a traffic accident or horoscopes concerning the upshot of battles and expeditions.

In complex causal systems – characteristic for the explanation of events that involve human action – such pseudo-explanations, however, have at least one true aspect. A driver, for instance, who believes in astrology can be insecure and this insecurity may negatively influence his or her driving style. In complex causal systems the explanation presupposes a causal analysis that makes a difference between primary and secondary, ultimate and proximate causes. Complex explanations, then, consist of propositions about causal chains or causal hierarchies. This way the notion of explanation, which refers mainly to events, is extended to an explanation of explanations. That means that according to their logical structure explanations basically allow for an infinite process. Hence, an explanation, especially a scientific one, cannot represent a definite explanation in an absolute sense. Rather, as will be shown, the hierarchical structure of explanation represents the real development of empirical sciences in a scale of lower and higher degrees of abstraction. For example, Newton's historically younger general law of gravitation represents an explanation of Galileo's laws of falling bodies as well as Kepler's laws of planetary motion.

One has to consider that hierarchies of explanation are always directed towards the very cause of an event, so that the notion of explanation in its narrower sense remains restricted to causal explanation. With the extension of the concept of explanation to human action, however, there arise problems.

Here the notion of explanation is in conflict with the notion of understanding. An example given by Rescher (1967, pp. 215-219), can help us make the problem clear. Let's suppose, a man plants his fist on the table:

- This action can indeed find a causal explanation: The man is "furious" or has had one drink too much. Such statements give particular answers to the question "Why has he done it?" They answer the question according to the modus of causality. Significantly, they do not refer to the decisions of the actor, but to the effect of impersonal "forces".
- If one sees the man's action as a result of his conscious decision, then there are other questions to be considered, e.g. "What is the goal he has been aiming at?" "What has led him to this action?" By answering these questions – e.g. "He wanted to frighten his interlocutor" – we intend an understanding of the motives and purposes behind the action.

There follows that the notion of action includes two kinds of elements: outer (physical, observable) elements and inner (mental, non-observable) ones. The term "action" always presupposes conscious decisions.

Merely causally explainable types of behaviour, like instinctive behaviour or reflexes (in humans as well as other living beings) are not to be considered as "actions" in a strict sense.

This difference is relevant to juridical methodology: As the definition of "complete description" offers rules for the (juridical) factual statement, the definition of "causal explanation" is linked to the question of guilt. The old Greek notion *aitia* offers evidence for this connection since it means both "cause" and "guilt".

Indeed, in ancient times the behaviour of animals and plants and even non-living objects was treated like human action. In Athens there was a special court where legal proceedings could take place against a stone, a spear, or any other object, which had incidentally killed a human being. In the Middle Ages it was still possible to bring an animal to the court and to sentence it like a (human) criminal (See Amira, 1891, Kelsen 1977, p. 32).

1.2.3. *Understanding*

Irrespective of its demarcation from "explanation" the concept of understanding is still ambiguous. Primarily, it refers to the mentioned "inner", i.e. mental aspect of human action. But since this cannot be observed, there are only physical *expressions* of an action that are the real, perceivable point of reference. The process of understanding is the critical reflection of the sense or meaning of those expressions.

Physical expressions of an action are not only those accompanying the action, so to speak, but also all expressions that appear at the end of the action as its results, consequences or products. Therefore – without leaving the initial framework, the conscious human action – we can apply the concept of understanding also to "understanding" a piece of art, a picture or a novel. This way it gets its particular meaning as "interpretation". The general concept of understanding refers to all kinds of signs and symbols as expressions of psychic experience. The concept of interpretation, however, is restricted to a limited category of signs, namely written documents including all types of documents and monuments "fixed" in a way similar to writing. Thus, the interpretation of a text becomes the paradigm for understanding as a method of the humanities in contrast to scientific causal explanation.

This contrast is constituted only on the basis of a methodological differentiation separating "explanation" from "understanding" at the level of everyday knowledge. Originally, however, explanation and understanding mean rather the same. Like any explanation, the process of understanding reduces the unknown to the known. Only the modus is different. An explanation reduces an event to a real "cause" or an "objective" law that encompasses all cases of the same type. This means that the event in question is considered as an element of a class of events. Yet, understanding a human action means to bring its expressions into a general system of sense. This system is known in advance, because there is the "essential congruence and interrelationship" (Gadamer, 1965, p. 204) between author and interpreter – any single human being is connected with other such beings and his or her community through "language", "religion", "legal systems", and so on.

While explanation is characterized by objectivity, understanding is subjective and refers to the psychological category of empathy, which is linked to *Verstehen* in its original sense. The subjective character of understanding leads to another difference between this method and "explanation". In fact, an explanation brings basically nothing new, whereas understanding has a heuristic relevance because it can reproduce *and* change or extend the given context. Yet, this presupposes an uncertainty of the contextual meaning, which leaves space for creative freedom. A particular kind of understanding or interpretation follows from notions that do not have just descriptive but prescriptive or normative functions. Any norm or rule can be followed only if it is *understood*, i.e. if one is able to see it in a particular context of sense. This context, however, requires – in the case of each application of the norm – an interpretation. The interpretation causes the free or at least conscious – and not forced or automated – compliance with the norm. In everyday life this happens intuitively because the particular norm is related to a known purpose or to a particular community, family, state, religion, etc. with a specific corpus of rules. If one is unable to grasp intuitively the sense of the norm, then he or she requires an explication or wants to understand the motives of the real or fictional person or institution establishing the norm. An example in juridical terminology is the well-known question concerning the "will of the legislator".

This general distinction between concepts with descriptive and prescriptive character gives us the basis to the methodological differentiation between juridical interpretations as normative hermeneutics and the humanities' interpretation as non-normative hermeneutics.

1.3. *Structure of Scientific Knowledge*

What are the general criteria of what is called "scientific knowledge?" Because of the permanent changes and different contents this question can be answered only formally or structurally.

The most elementary structural feature of science is its systemic character. Unlike pre-scientific, common-sense knowledge, scientific knowledge displays an objective system-related structure of order. This order can be related to

- The static aspect, i. e. the context of the products of knowledge which yields the definition of science as a logically "closed" system of concepts and propositions;
- The dynamical aspect, i. e. the process of gaining, extending and changing knowledge that leads to the definition of science as a *research*, a methodologically organized activity or action based on the concept of "open system".

The difference between "closed" and "open" systems plays a central role in today's juridical methodology:

- The concept "closed system" was coined by Heck, the main advocate of "jurisprudence of interest", in order to characterize the theoretical construction of conceptual jurisprudence (Heck, 1932, pp. 116-118, Esser 1964, p. 222).
- The distinction between "open" and "closed" systems has been used in the context of Roman legal history by Fritz Schulz and has found its relevance in comparative law studies (Schulz, 1934, p. 84, pp. 362-364, Schulz, 1946).

- The methodological characterization of the "open" or "dynamical" system is the basis of the hermeneutical method and is not influenced by the very definition of hermeneutics, either as a method of interpretation (Larenz) or as a method of application and finding law (Esser) (See Canaris, 1969).

1.3.1. *Scientific Concept Formation*

When looking for basic elements in the structure of science, we find in first instance the formation of concepts. Primarily, each scientific discipline is a conceptual system.

The classical ideal of the logically ordered conceptual system is the so-called "pyramid of concepts" that makes possible to start at any point and to go through a series of intermediate links up to the most general concept that includes all the other concepts. Only when scientific concepts, defined in this manner, represent a systematic unit, the respective scientific discipline can be formulated as an axiomatic-deductive system of propositions. Thus, the decision about this possibility is already made in the process of concept formation. However, this decision refers exclusively to the formal logical structure and not the contents of concepts. Only unambiguous concepts can build the fundament of such logically correct systematisations. Unequivocal concepts do not just determine a clear definition of the extent of a concept in contrast to other concepts, but are also related to the clear decision concerning the question whether or not an object is characterized by a concept.

From this follows the characterization of a concept in terms of "logic of classes": Concept, then, is nothing else but the name of a class of objects.

This view, however, is formal logical and not epistemological, because the concept of classes can be defined in whatever way, independently of the real existence of objects. An evidence for this is the notion of "empty class" which is used in modern logic. The classical example here is the unicorn. It is defined as an animal looking like a horse, but possessing a horn on its forehead. This is indeed a reasonable definition giving sense to the term "unicorn". It defines a class. This is not a class useful in zoology for it is empirically empty – it has no elements – but this is not a logical proposition (Carnap, 1974, 65).

If in the process of abstraction we do pay attention to the negative aspect of "omission" then we see that each decrease of conceptual content increases or extends the scope of a concept. A concept loses its content with increasing generality.

If in the process of abstraction we do pay attention to the positive aspect of displaying essential characters then by a concept – irrespective of its generality – we have to understand a complex of characters. The concept is more "concrete" when it does fit reality.

The distinction between "abstract" concepts of classes and "concrete" concepts is a central issue in contemporary juridical methodology, which is directly connected with the distinction between "closed" and "open" systems:

- The advocates of a "juristic logic" rely on the ideal of a logically "closed" axiomatic-deductive system and always adhere to the exactly defined concept of classes.
- Those who advocate the "open" system support at the same time the idea of a "concrete" concept.

Yet while logic of classes and the idea of axiomatisation are unequivocal programs in philosophy of science, the doctrine of the concrete concept is as ambiguous as the conception of "open systems". Thus the "idea of concretisation" (Engisch), which is based on concepts at the "medium level of abstraction" is to be separated from the view of the "concrete" concept (Larenz) which has its roots in Hegel's absolute distinction between concrete-general and abstract-general concepts.

The ambiguity of the doctrine of the "concrete concept" is based on the fact that it can have different contents leading to different conceptual forms that – with regard to contents – are determined by particular areas of objects. The classification of concepts, i.e. their arrangement as conceptual forms, is therefore to be understood as a differentiation of the "concrete concept". This classification is related to the ways of representing knowledge contents. Since jurisprudence refers to all objects of reality, which can be legally regulated, it includes basically all conceptual forms, which have been systematized in philosophy of science as follows.

The descriptive-representing concepts are the most concrete concepts that build the starting point of all scientific knowledge. They are very much constrained by the given situation that means that outside the concrete context of the objects they describe there is no possibility of further elaboration. The basic principle of their accomplishment is the similarity between the representation of facts and events. They therefore most clearly appear in the so-called "protocol statements" which serve the ideal of the "simple representation", i.e. the pre-theoretical, pre-systematic description.

The development of these concepts into classifying concepts does not mean any gain of information, but just the generalization of a particular aspect of gaining an abstract order, which by itself serves only a diagnostic purpose – the recognition and arrangement of the "same" objects and facts according to certain criteria. Unlike descriptive-representing concepts, the classifying concepts are already general conceptual hypotheses that allow for incorporating the same object into different systems of order.

The jurist is not interested in the zoological classification and differences between mammals, fish, etc. but rather makes differences between domestic, wild, and tame animals. This differentiation is zoologically irrelevant and based on the relation between animals and humans. A jurist thinks in terms of the "external system" and does not classify an "animal" as a "living being", but as a moving object. The reason for this is that law includes many norms concerning such objects, e.g. acquisition or loss of property. Virtually, all these norms can be applied to animals (Larenz, 1979, p. 433).

The comparative typological or topological concepts do not merely aim at representation, but at comparing real phenomena. Like the classifying concepts and unlike the representing concepts, they are therefore already general in character. While, however, the classifying concepts are only names for unarranged sets, the comparative concepts "reproduce" more or less ordered structures.:

- The concept of type in comparative anatomy in the sense of Cuvier represents the *Bauplan* ("design").
- The group of elements or the periodic law in chemistry represents a topological structure according to which chemical substances occupy a particular place in the system.
- The typological concepts in jurisprudence represent the structure of "mutually related regulation complexes the elements of which are contents of norms together with particular relations to life" and are "connecting links" between abstract generality and individuality (Engisch, 1968, pp. 259-260, Larenz, 1979, p. 445).

The quantitative and metric concepts are those that – in the sense of logic of classes – are so precisely definable that they meet the standards of the strict logical equivalence or identity. Only at this stage abstraction develops into a concept-formation processing in the sense of basic logical operations. The fundament of these concepts is a strictly defined metric system.

Quantitative and metric concept formation is to be found mainly in the exact (natural) sciences; their objects can be counted and measured. However, the tendency toward quantification is traced in all areas of science. In jurisprudence, too, metric concepts are used in all those cases where technological aspects are subsumed under legal rules. For example, a person can deal with radioactive substances if he/she is legally permitted to do so or if he/she is acting in particularly defined areas that are subject to regulation of radiological protection, which includes a strict definition of radioactive substances (See Raisch, 1970, pp. 434-436).

The system of scientific concepts as far as it concerns empirical ones is completed with the qualitative concepts (as descriptive and comparative concepts) and the quantitative and metric concepts. Basically, the development of practically each single scientific discipline includes, potentially or actually, all these concepts. Hence, the structural differentiation of concept types does not yield an absolute criterion of classifying or evaluating a system of the sciences, i.e. in the sense of distinguishing between exact and non-exact sciences. Rather, the use of certain forms of concepts is strongly connected with the given field of objects and its particular problems. The same is true to the so-called theoretical concepts. The sense and meaning of these concepts follows only from the complete system of scientific theories. They are in themselves very vague and poor with regard to contents, without any special information, i.e. concepts such as "space", "time", "motion", "life", "law", "state", and "property". In the context of a scientific theory, however, they gain the status of fundamental concepts with the highest degree of information condensation.

Thus, there are different but complementary structures of order, according to which conceptual systems are arranged. This means that each concrete content-determined concept can be basically represented by a class concept, if this is required by the formal-logical order of the scientific concept system. This order disregards the specific contents; the latter has, however, to be considered again in the case of the concrete application of scientific concepts. This is true to each type of scientific knowledge, no matter whether this concretisation is accomplished by the "hermeneutical interpretation" (as it is the case with the humanities) or by quantitative and metric representation of empirical data (in the so-called exact natural sciences). Like all sciences, legal science is characterized by a specific form of concepts, although - as the law of radiological protection shows - it can even include quantitative and metric concepts, because physical and technological facts, too, can require legal regulation. However, in relation to the legal system such concepts are external concepts. The internal juristic concepts expressing legal conditions are frequently called "typological concepts" (See Engisch, 1968, pp. 237-239, Larenz, 1979, pp. 443-445, Leenen, 1971). An example are the "types of contracts" in the *Bürgerliches Gesetzbuch* (Civil Code) that according to Larenz are "true types" in the sense of fuzzy concepts displaying "mixed forms".

Yet there is still no agreement about the relation between classifying and typological concept formation within juristic methodology. Carl Hempel's and Paul Oppenheim's *Der Typusbegriff im Lichte der neuen Logik* (1936) (The Concept of Type in the Light of Modern Logic), however, is regarded as "most important" (Engisch) also in juristic methodology. This publication elaborates the difference between classifying concepts that connect or disconnect objects with particular properties, and the typological concepts that, according to a hierarchy of properties, place these in the system. A vast number of juristic publications concerning the type concepts deals affirmatively or critically with this study (Kuhlen, 1977).

1.3.2. *Scientific Hypotheses and Theories: Static Structure of Science*

In first instance, science is to be regarded as a conceptual system. However, the primary task of the concepts is, in principle, the formulation of statements about a particular field of objects. The most simple kind of statements expresses the existence of certain facts by means of empirical, qualitative and quantitative concepts.

In addition to the protocol statement as an empirical statement about facts, there is another type of statements: the "universal statement" (*Allaussage*) which does not refer only to one or more objects or facts, but to the complete object area of a given field of scientific research.

The single and particular statements in all sciences serve the purpose of description. Hence they include descriptive concepts or – since comparison and measurement are the continuation of the common, qualitative description – comparative and quantitative and metric concepts. Universal statements, however, have another function. They do not refer only to the already recognized sphere of empirical objects, but transgress this sphere and are related to all those objects, facts, and events that will be empirically grasped in the future. Epistemologically, therefore, universal statements have to be considered as hypotheses.

Yet one has to make a distinction between mere generalizations (e.g. "all swans are white") and universal hypothetical laws (e.g. Kepler's first law of planetary motion "Planetary motion is elliptic").

This distinction is epistemological by nature, because from the point of view of formal logic law hypotheses do not differ greatly from mere generalizations: In both cases one uses words like "all" or "every". Epistemologically, however, mere generalizations have the character of an accumulation of single recognitions that, in the best case, can be extrapolated as expectations concerning future events, while universal law statements display an explanatory and prognostic character.

According to their logical structure, explanation and prognosis are not different. Both have the form of a deductive conclusion:

- Major premise: Law hypotheses as general statements.
- Minor premise: Special conditions as special statements.
- Conclusion: The explanandum or event that has to be foreseen (See Hempel, 1977).

This simple logical structure alone shows that the explanatory and prognostic function of the law hypothesis can be accomplished only within a wider complex of statements.

To characterize legal science, we can still refer to the already quoted remark by Engisch: "Like natural science legal science is a science of law" (1971, p. 9). But the notion of law in legal science differs from the law concept in natural science and demonstrates ambiguity thus causing controversies:

- In the strict normative conceptions law is identical with the pure norm stated as a pure "ought" independent of legal reality, i.e. human behaviour. In this sense, law has not a hypothetical, but an absolute character in a timeless legal system.
- However if one takes into account that law, as positive law, changes according to the social constraints and requirements, then the juridical concept of law is just a "normative expectation of behavior" (See Luhman, 1972, p. 105, Raisch, 1973, p. 27), has hypothetical character, is amenable to "inductive" control, and undergoes changes.

The condensation of general and particular statements in a non-contradictory system is called theory. As a system of statements the theory's logical structure consists of two sets of classes:

- Axioms or principles
- Theorems or derived propositions

If, and only if, concepts are determined as unequivocally defined class concepts, then they fit the axiomatic-deductive representation as fundamental and derivative concepts in a theory. However, such unequivocal concepts exist only in a few disciplines, e.g., mathematics or the so-called exact natural sciences where concepts are determined with respect to their quantitative characteristics as classes of equivalence. Nevertheless, one speaks of theories in many other areas of science let's just think of economic theory, theories of state and society, and legal theory.

Whereas theory in the (natural) sciences is a means of construction that helps the process of explaining and predicting by unifying experiences, those theories are directed towards understanding human action as a realization of intentions and norms. Society, economy, state, and law do not represent external objects like the objects in the natural sciences. These areas do not remain untouched by theoretical considerations. Unlike explanatory theory, the "understanding theory" changes its objects, because it does not simply refer to a given order but is rather participating in the formation of ordered systems (See Gadamer, 1965, p. 431).

In this sense, social and economic theories are not just "cognitive" or "contemplative" theories. Information concerning lawfulness leads to a process of reflection that may change both the special, empirically understandable conditions of law and law itself (See Habermas, 1968).

Legal theory has a particular position. Its object is law. "Law" can be understood in two different ways that are clearly indicated by the two concepts *ius* and *lex*:

- "Law-in-itself" which has always been an object in legal philosophy, e. g. as "natural law" or the "true law" that has in fact its history, but does not coincide with its real historical figures (See Kaufmann 1971: 81ff).
- Law as the totality of (legal) prescriptions in any given legal system is a subject matter to legal dogmatic (See Kaufmann, 1962, pp. 397-399).

As far as law represents an order created by the human mind, it is already a conceptual system and a system of sentences by itself, dominated by logical rules of connection. The difference between these systems and cognitive systems, which, as empirical theories, give information about the real world, is – in a formal logical sense – the use of non-descriptive, but prescriptive language. Legal norms by themselves are indeed no sentences or judgements concerning a given object, but restrictions, permissions, and authorizations. None the less, legal theory cannot be simply reduced to the logic of norms (See Klug, 1966), which, as a formal structural theory of law, refers to the static of law.

Already Kelsen emphasized that one has to make a distinction between a statistical and a dynamical legal theory; the latter deals with law in flux, so to speak, with the legal process, which is creating and applying law (Kelsen, 1960, p. 72). Independent of the position of legal positivism which forms the background to Kelsen's *Reine Rechtslehre,* (Pure Theory of Law) this insight (that law regulates its own product and application) has been expressed in the whole area of juristic methodology. As an "understanding theory" legal theory is not a formal logical reconstruction or an "*a posteriori* correction" of the structure of law as a system of norms, but, in first instance, a "critical reflection" regarding law not as an "object", but rather as a "result" of this reflecting knowledge process (See Kaufmann, 1971, p. 3). Thus legal theory

– also concerning its own contents – transgresses legal dogmatic which does not aim at "justice" but at "legal security".

In methodological terms this means that legal theory is primarily acting as a meta-theory of legislation and decision-making (See Kaufmann, 1971, p. 27). Here we can already notice a solution of the classical methodological controversy. This solution has been found independently by the "scientistic" analysts and the advocates of hermeneutics oriented towards the humanities and is the basis of the evolutionary view of law. One can speak of self-regulation or self-reflection of law, but in both cases the intended legal theory conceives of law as an open, self-developing system with changing basic principles according to an ever-more complex legal reality. The presentation of the "evolution of law" does not consist only of a mere description of changes.

The biological theory of evolution, too, goes far beyond describing the modification of living beings. It *explains* these modifications by the means of particular mechanisms. Accordingly, we have to find a mechanism in the sphere of the "evolution of law" in order to explain the changes of legal systems. Such a mechanism is to be found in the functional context of scientific method that exists in all fields of the sciences, in the explaining (natural) sciences as well as in the "understanding" humanities and in applied sciences like medicine and jurisprudence.

Philosophy of science, then, has not only to deal with the static structure of theories but also with the dynamical aspects of science as a research, i.e. with problems of the formation of hypotheses and theories and with the problem of modification and replacement of theories.

1.3.3. *Dynamics of Theories: Functional and Dynamic Context of Scientific Methods*

The operation of building hypotheses has been at all times *induction.* According to its logical structure, this is the step from particular statements to a general statement. Logically, however, such an inference is not valid. Unlike deductive inference, the step from a general statement to particular statements, induction does not give any guarantee for truth.

Observations of white swans do not justify the conclusion that all swans are white (After Popper, 1973, p. 3). Nevertheless, any scientific discipline includes general statements. Nay, the formulation of such statements as hypotheses is the outstanding criterion of scientific knowledge. Also, the true sense of inductive inference is not given by its formal logical structure, but by the epistemological function of the extension of our expectation beyond direct experience.

This way inductive logic develops into a pragmatic rational decision theory (See Stegmüller, 1974, Oeser 1976, pp. 38-40), which is not to be understood formally theoretical, and the inductively obtained statement reveals as a basically subjective hypothesis with a certain amount of risk. Each universal statement, as far as it is not a tautology, is hypothetical in character. It is valid only as long as it is corroborated again and again by experience. This shows that there is a second kind of induction. It comes at the end of a knowledge process as its test, and is not "leading" particular statements to a general statement, but rather "reducing" the universal, already proposed as hypothesis, to the particular empirical statement. This procedure can also be understood as quasi-induction, i.e. a deductive conclusion towards induction, because it does not use the special empirical statements for gaining the universal statement, but starts with these statements in order to make a conclusion about truth or falsity of the general statement.

Ever since the establishment of the modern sciences, the two kinds of induction have been known as

- Enumerative induction, and
- Eliminative or excluding induction.

The problem of induction itself that includes the question of the acquisition of general hypotheses as well as their justification can be solved only by a systematic connection of both types of induction in a functional model of the whole knowledge process. It is visible that the simple enumerative induction by no means determines the process of building universal law hypotheses, because a mere enumeration leads just to generalization (remember Popper's example). However, law hypotheses like e. g. Kepler's laws of planetary motion are not based on a simple enumeration of data, but a "saltation" that happens by the discovery of the super-concept of the law statement, e. g. the concept "ellipse" in Kepler's first law. This creative "super-induction" that facilitates the saltation from particular statements to general law statements is therefore directly connected with the abstraction of the procedure of concept formation.

- In the case of concepts representing according to definition class concepts, such law hypotheses can form a system in the sense of an axiomatic-deductive theory.
- In the case of not exactly defined concepts – like the typological concepts – one can indeed construct a system but its particular statements (in their

own context) cannot be simply inferred. A special proof is required to show whether or not they belong to the system.

If we have to summarise the general logical and epistemological methods according to their functional correlation together, we would attain at the following model:

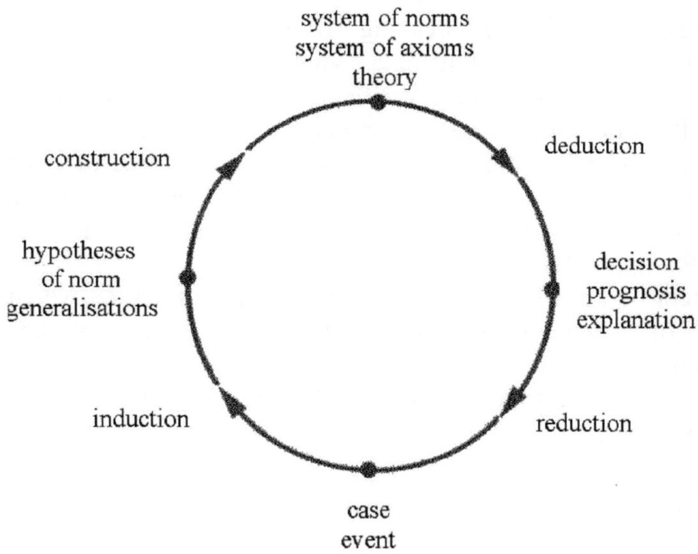

This general model of the scientific knowledge process applies for:

- The feedback process of theory and experience in the natural sciences,
- The so-called hermeneutical circle in the humanities,
- And finally the "applicative" circle of making and applying law in jurisprudence.

All such models are based on the functional context of induction and deduction, which had been already known in Greek philosophy (Plato, Aristotle) as the "way up to the principles" and the "way from the principles down to the singular statements".

Galileo and Newton introduced this model of the opposite but inseparably connected operations to the (natural) sciences. In the humanities Dilthey presented it as a circle of understanding, a "cooperation of induction, application of general truth to the special case, and comparative proceedings". The applicative circle of jurisprudence already follows from the peculiarity of juridical hermeneutics. This way the problem of induction becomes more serious: a leap not only from "special or singular statement" to "universal statement", but also from *is* to *ought*. The justification of using "induction" in legal science (Esser) is shown by those "norm hypotheses" (Kriele) at the middle level of abstraction that create in each juridical decision directly or indirectly a mediation between abstract norms and concrete cases. Hence, it becomes apparent that inductive inference is a primarily pragmatic operation allowing for "breaking off" a theoretically infinite argumentation. The "breaking off" refers to the heuristic gaining as well as the a posteriori testing of a hypothesis that – even in the natural sciences – is neither absolutely verifiable nor absolutely falsifiable. In the field of all empirical sciences the process of examining a concrete case comes to an end when the corroboration is tolerably free of errors. In the juristic sphere the process of finding law often ends with a result to the best of one's knowledge and belief because one is forced to come to a decision (Kriele, 1967, pp. 191-193).

All scientific knowledge has basically a similar structure, which justifies (the unity of) a general concept of science. The variety of the complex phenomenon *science* in its real development is based on the fact that this (unified) structure displays varieties of weighting and meaning of the methods it includes.

1.4. *Classification of Sciences and the Position of Jurisprudence*

In juridical methodology the problem of classification of sciences has always played a decisive role – understandably, because the problem of the concept of science in jurisprudence, which, as I have already shown, is the starting point of all methodological reflection, can be solved, after all, only if the position of jurisprudence in the system of the sciences becomes clear.

In this chapter I discuss the problem of the classification system of sciences from the point of view of contemporary philosophy of science. Also, I explain the present modified situation of jurisprudence as a neighbour of social sciences (See Grimm, 1976).

Ever since its beginnings as an autonomous discipline in the 19th century the task of philosophy of science has been twofold:

- It has to represent the methods that bring forth scientific knowledge and according to which science is taught.
- It has to give a systematic foundation of the "natural classification" of scientific knowledge (See Ampére, 1834-43).

Both tasks are inseparably linked together for each classification system of the sciences has to take into account the methodological perspective and, *vice versa*, each specialized methodology that is expected to do justice to the pluralism of methods in the sciences starts with a classification system. This interrelation is not circular but rather represents the modus of the origins and developmental structure of the system of sciences. As we have seen previously, the knowledge operations already applied to our everyday life – description, explanation, and understanding – are the fundament of the systematisation of science. Therefore the methodological aspect is prior to all criteria of classification. According to different aspects an object can be subject matter to different areas of science.

The "natural order" of objects, then, cannot give us a clear arrangement of the sciences. For this reason in practically all our attempts to classify the sciences we have used primarily aspects that refer to the modes of recognizing objects and not the objects themselves.

Sophisticated classification systems as they were developed already in the 19th century were based on criteria of classification that refer to the method as well as the result and its application in the form of contrasting concepts like

- Simple – complexe (Comte),
- Abstract – concrete (Spencer),
- Theoretical – practical, applied (Ampére, Cournot).

The many still relevant attempts notwithstanding, there is today no generally accepted classification system of the sciences.

The reasons for the difficulty or impossibility of such a system are the following:

- There is a variety of contradicting perspectives for the classification of the sciences.
- Sciences develop quickly, new disciplines emerge, known disciplines disappear.
- The attempt to construct an objective, "natural" system of sciences is circular: A "natural" system includes the claim for a classification, which truly represents the real structure of the world. This presupposes a world model, which, however, would be the result of science and its underlying scheme of classification.

These considerations contradict any attempt to establish a definite, unchangeable, static system of the sciences, which would do justice to all criteria of classification and comprise all disciplines according to their specific differentiation. But, at the same time, these considerations do not oppose the *possibility* of representing the phenomenon of science in a comprehensive system of order, which can offer the framework to any further differentiation. The conditions for such a system of order are the following:

- It has to include the structures of order that were already known and useful in the past,
- It has to meet the requirements of today's science and consider the present status of the sciences,
- It should not display a logically closed structure, but leave space for changes.

This way we can also get rid of the objection that the system is circular, since this objection refers only to a statical, hierarchically built system. The "natural" system of the sciences becomes a "living whole" (Oppenheim, 1926, 1) its context does not only represent logical relations, but is the result of a real historical development that follows certain dynamical laws.

These laws are constituted by the functional relations among the fundamental types of methods. According to these laws the system of sciences is originally arranged and developing like an organ. Hence, we get here a topological, i.e. non-linear and not hierarchically arranged model of order. In this model the position of each single scientific discipline is characterized by its methodological genealogy.

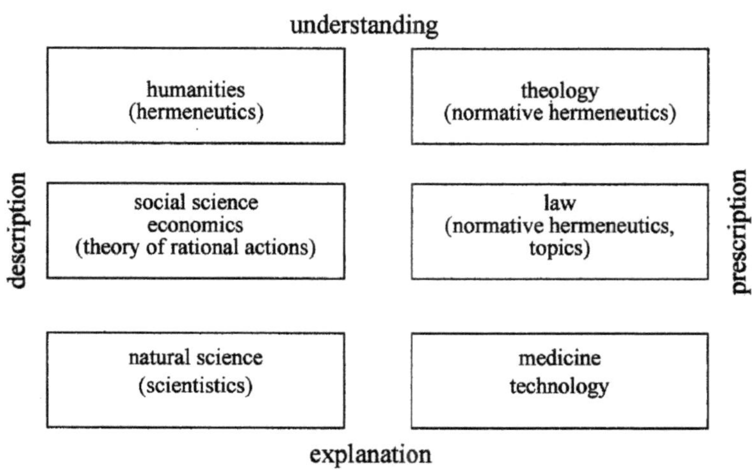

Let me explain. The category of description is included in all fields of science, because each discipline refers either to the description of empirical objects in nature, or the description of human action (or the results and expressions of this action). The prescriptive component is also penetrating, as it were, all spheres of science, but, unlike description, is prevalent in sciences that are connected with purpose and application. In this sphere – which represents the oldest corpus of human knowledge – the extra-scientific origins of science are most evident. The three faculties that have been the essence of the European University since the 12th century were systematically and reasonable characterized by Kant in the following way:

- Theological Faculty – external welfare
- Juridical Faculty – civil welfare
- Medical Faculty – physical welfare (See Kant 1968).

Purely descriptive knowledge is extended by the categories of explanation and understanding. However, this does not create a dichotomy but rather there is a third field in which the transition of explanation and understanding becomes systematically evident. The objects of this field are human activities with their different individual and social relations. The field was systematized quite late; remember that psychology and sociology appeared only in the 19th century. The same is true to economics. So, one can understand the traditional scheme of sciences which includes just natural sciences and humanities. Only the empirical method separated economics and social sciences from the humanities in a narrower sense.

The scheme of three spheres, too, does not represent a completely differentiated classification system, but just a basic model from which further differentiation and gradual transitions can be derived.

In contemporary philosophy of science the most common attempts to classify science follow one of three lines and always include three complexes:

- Oriented towards objects, like Popper's theory of the three worlds (nature, human mind, products of human mind) (see Popper 1974),
- Structured according to knowledge interest, like Habermas' approach (empirical-analytical natural science, historical-hermeneutical

humanities, and artificial-systematic sciences of action) (See Habermas, 1968, p. 155-157).
- According to fields of objects and methods, like Diemer's model (metric sciences, behaviour sciences, and sciences of meaning) (See Diemer, 1975, p. 7).

Finally, the partition of sciences into three main fields finds its manifestation also in "general philosophy of science" which should not only include (natural) sciences, but also humanities and social sciences. There is strong evidence for this partition in the extension of the concept "meta-theory of science" by Stegmüller who proposed the following three concepts: "meta-theory of science", "meta-theory of humanities", and "meta-theory of rational action" (See Stegmüller, 1974, p. 8).

This characterization of the general concept of science helps us recognize the three fundamental (and only in their extremes contradictory) programs determining the present situation of philosophy of science.

The reductionist program (language analysis), also to be characterized as scientistic, reduces science to the methodological ideal of exact sciences. It was originated in the older positivism of Comte who had been fascinated by the progress of mathematics and physics since Newton. This positivist view was continued in so-called neo-positivism of the Vienna Circle, which separated from the old sensualist positivism and developed into the so-called analytical philosophy of science ("logical positivism", "logical empirism"). However, according to its self-conception analytical philosophy of science is no longer "positivistic" in any sense, but rather neutral. It has reduced science to a "system of statement" and is not restricted by empirical science. This reduction is possible, because analytical philosophy of science is to be understood as "logical analysis of the language of science". The advantage of this program is the precise elaboration of the formal structures of argumentation that allow for constructing science as a consistent (i.e. logically correct) system of statements. The model was meta-mathematics or the meta-mathematical theory of evidence dealing with the proof that axiomatic-deductive systems are free of any contradiction. From this point of view, a meta-theory of the empirical sciences was intended which should also solve the problems of empirical corroboration of (empirical) theories.

The disadvantage of this reductionistic, i.e. formalist program is its drastic limitation of problems posed by philosophers of science. The program excludes all those sciences that, because of their complex objects, cannot meet the standards of an axiomatic-deductive system of statements, i.e. the humanities as well as the so-called descriptive and comparative (natural)sciences. Moreover, the program also largely excludes that area of problems that has always been at the basis of problems with regard to theories of evidence and corroboration: *heuristics* or the study of finding or gaining knowledge in science.

The explicit expression of this fundamental reduction of the questions of philosophy of science is the dogma of the separation of the "context of discovery from the context of justification" (See Reichenbach, 1938, p. 7-8).

The dogma tells us that a particular part of scientific knowledge, namely creative thought leading to the recognition of general laws, is not amenable to any rational analysis. It finds its methodological expression in the so-called problem of induction that, with regard to its formulation according to logic of propositions, represents an unsolved dilemma (See Lenk, 1971):

- Inductive inference leading from special to general statements is indeed an extension of knowledge, but does not save the truth, i.e. it is not logically valid;
- Deductive inference leading from general to special statements is logically valid, but not extending knowledge, because it implies the already known.

Yet another reductionistic consequence is connected with the unsolvable problem of induction in the analytical, formalist philosophy of science. For the emergence of anything new cannot be rationally explained, the progress or development of science, too, is not explainable. Between preceding and successor theories there is no relation that could be represented as a formal-logical relation (thesis of incommensurability).

However, the solution of the problem is, from the point of view of analytical philosophy, that one has to leave the original program of the "logical analysis of the language of science". Rational theory of decision, then, replaces the problem of induction, and the conception of science as a system of statements is replaced by a dynamical structure including a structural frame, a structural nucleus, and an extended nucleus. This way, after all, from the point of view of model theory, science is to be characterized as a topological system of concepts.

The hermeneutical program is not only the elaboration of an alternative, but its advocates have always intended a universal claim. It is rooted in 19th century methodology of the humanities (Dilthey) and can be traced back to the "philosophy of mind" of German idealism, especially to Schleiermacher (See Gadamer, 1965, p. 162-164).

However, the concept of hermeneutics is much older. It was already used by Plato and Aristotle and its original meaning was "the art of interpreting (texts)". According to different texts, already in ancient times hermeneutics found different fields of application:

- The interpretation of works of fiction leads to philosophical hermeneutics.
- The interpretation of Jewish, Christian, and Islamic religion constitutes the sphere of theological hermeneutics.
- The interpretation of juridical texts brings forth juristic hermeneutics.

Only Schleiermacher elaborated a universal hermeneutics. Dilthey approached hermeneutics as a methodology of the humanities by adding Droysen's "historics" as historical hermeneutics.

The universal claim of hermeneutics was innovated by Gadamer (1965). In a "philosophical sophistication" he tried to show that hermeneutical methodology was not to be understood as emerging from the competition between explanation and understanding. Rather it is a consequence of meta-theoretical communication referring to the historically caused "world understanding" which includes the world model of explanatory (natural) science. This claim of hermeneutics as methodology of understanding is justified, if it is understood as quasi-circular fundamental structure of human communication and tradition. In this sense, hermeneutics is not different from the reductive structure of explaining explanations (in the natural sciences) through which the historical process of the development of scientific theories is to be characterized.

The action-theoretical program has been mainly developed in social sciences, but – in its wider sense – it has originated from an operational and instrumentalist view of science itself. Thus, intuitionist constructivism has characterized mathematics as an action, and theories of the (natural) sciences as instruments not of knowledge, but of mastering nature. Already in this sphere there is open structure of a self-correcting system in the process of the application of scientific knowledge as action or activity. Independent of this functional mechanism of information gaining, which is related to human knowledge, the structure of self-regulating processes has been recognized in technology (cybernetics) and in biology (system theory). So the intuitive conceptions of the human knowledge process have become clearer (See Oeser, 1976). System-theoretical considerations go beyond studying functional mechanisms of human knowledge in the area of (natural) sciences; they have been also developed in the field of the sciences of action as a special variety of "understanding sociology" – the so-called functionalism (See Bühl, 1975) – and in economics as *game theory* (See Neumann, Morgenstern, 1961).

In their mutual relationship, these three basic programs of philosophy of science are a unified complex of general methodological rules that in their problem-oriented differentiation, form the fundament of all special methodologies of scientific knowledge.

Juristic methodology itself gives in its present status an example of the necessary integration of philosophy of science. Within its frame the three programs have been immanently developed:

- The analytical program (language analysis) through formalistic logic of norms (Fiedler, Klug, Tammelo) was further developed in the analytical theory of discourse that, however, has to be distinguished from the practical discourse theory, which transgresses analytical philosophy of science and belongs to the sphere of a theory of practical reason.
- The hermeneutical program (Larenz, Canaris) in connection with topics (Viehweg, Engisch) most conveniently supports the relation to an ontological position in legal philosophy.
- The action-theoretic program in its strict sense has not developed in the area of juridical logic of norms, but rather within the meta-theory of applying and creating law (Esser, Kriele, Fikentscher), and stands in connection with system-theoretical functionalism in legal sociology (Luhmann). This program does not transgress the dimension of juridical methodology as well – it leaves open the question of a more profound foundation of positive law.

1.5. *Model of the Functional Context of Juristic Methods*

Regarding its practical application, jurisprudence is different from the other sciences that have mainly to do with methods of knowledge. In Jurisprudence, "method" and "methodology" have a stricter meaning than in the humanities or in the (natural) sciences, because, according to the very task of legal systems, a more rigorous procedure of finding law is demanded. The historicity of such acknowledged ways of knowing and ways of founding for the solution of legal questions leave to juridical methodology a wider space – wider than in any other science that refers to areas of objects, which are independent of social conventions. This way, juridical methodology, together with juridical dogmatic, undergoes its changes. Such a dynamical juristic methodology comprising the historical changes of law, however, can be established within a comprehensive meta-theory of methods only if philosophy of science itself is no longer restricted to the logical analysis of static structures of scientific statement systems or hermeneutical interpretation models, but rather considers science as a conceptual system constantly in flux. Therefore, the axiomatic-deductive, i.e. static systems by which traditional legal philosophy has always tried to superimpose positive law is to be replaced by the notion of law as an "open" (Schulz, Esser), "movable" (Wilburg) developing system that could be best represented as a circle.

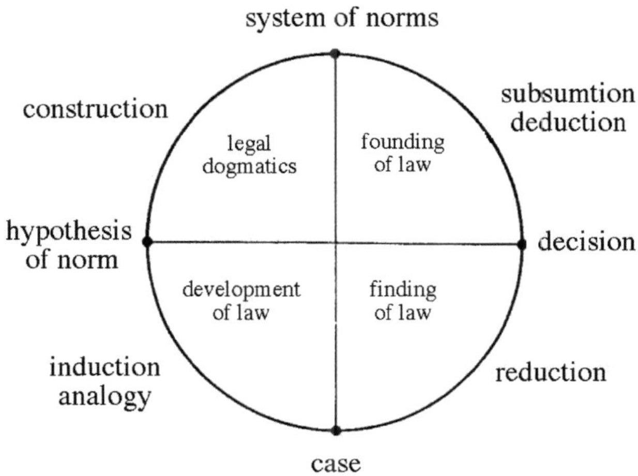

Let me explain the model. The upper half of this functional circle of juristic methodology represents law as a stabilized system of legal statements, i.e. norms or legal rules as they appear in the Statutebooks (e.g. Civil Code) of codified law. Methodologically, the question arises, how this system is structured. The general method of building the system is construction. The juridical construction refers primarily to codified law, i. e. a comprehensive code including the whole or a relevant part of positive law. In some rarer cases the juridical construction refers to the codification itself, that means to the act of making law like, for instance, the Justinian Codification or the establishment of the *Bürgerliches Gesetzbuch* (Civil Code) at the beginning of the 20th century. The juridical construction is therefore a kind of reconstruction of positive law in the sense of Radbruch, which brings the relations between the parts of the system into the consciousness. In this aspect of law as constructively founded system of normative legal statements all applications of law to concrete cases is a legal subsumption of special statements under the stated norms. Taking this aspect absolutely e.g. in concept jurisprudence, one restricts the respective method to grammar and logic, even if one does not start with the untenable presupposition that the codified positive law itself is already to be demonstrated as an axiomatic-deductive system of statements.

The lower half of the model is representing the concrete and practical sphere of applying law and the connected questions of decision making in accordance to cases and norms. This means the relation of the legal system with its internal juridical statements of norms (prescriptions, prohibitions, permissions, authorizations) to the empirical legal reality, which is expressed in the given concrete cases as empirical-descriptive statement. From the perspective of the empirically perceivable legal reality, application of law always means the search for a superimposed normative legal statement, which is already supposed to be known in the system of such statements.

One should note that in this model of functional relations between juristic methods a clear distinction *founding* and *finding* law is made:

- *Founding law* is the logical foundation according to the so-called syllogism of determining consequences of law. It belongs to the upper half of the model, which represents the systematic-structural side of law. This is the sphere of juristic logic in a narrower sense. Methodologically, this act of the logical foundation is always a deductive process. The system of law is, in its best case, quasi-axiomatic and because of the uncertainty of legal concepts it requires a "hermeneutical" interpretation.
- *Finding law*, however, is the constitution of a legal case by the determination of legally relevant aspects and given characters of the fact.
- *Gaining law* (after Kriele) is to be understood as the establishment of a normative hypothesis on the basis of given facts – "facts of life") (Kriele, 1967, pp. 197-199). It transgresses the application of law in the sense of founding and finding law only if the special normative hypothesis (helping the jurist to find the general statement for the concrete case) is not directly or explicitly included in the system of positive law, i. e. in the code, and also not covered by prejudice. Then it is necessary to rely on "extra-logical" fundaments of decision, which, however, are not subjective, but based on certain "standards" of legal thinking in a particular time (Esser, 1956, pp. 96-98). These extra-logical principles or examples are caused by law itself, especially by so-called general clauses.

In this model law is supposed to be a dynamical, developing system. According to the model also the extralegal basis of decision and evaluation – circumscribed as "legal-ethical principles" (Esser), "reason-legal argumentation" (Kriele), "nature of facts" (Kaufmann), etc. – has no static character, but is historically determined. It is not pre-given as a system of external values or perceivable by speculation, but rather originally discovered in the concrete case; this makes the inductive-heuristic component of the procedure evident. This process of gaining law takes place at the level of legislation as well as at the level of a judge's decision. It is neither purely inductive, nor merely deductive. "Laws can be made concrete only with reference to the facts of life which should be regulated. In the same way laws can be made real only when accounting for the real fact of life that has to be regulated" (Kaufmann, 1972, p. 28).

1.5.1. *Setting Law: The Static Structure of Legal Systems*

In the different legal systems of today, law is laid down in statute books as, e.g., in the German *Bürgerliches Gesetzbuch* or the *Handelsgesetzbuch* (Commercial Code) that together build a whole. On further consideration one notices that methodologically these are not axiomatic-deductive systems where one could find all types of behaviour that are relevant to law. It is apparent that the codes are the product of a multi-level developmental process. In many of their parts the Civil Code and the Commercial Code are "reproductions of Roman Law, however, modified and mixed with rules that were established in German speaking countries, and then codified in autonomous territories." (Raisch, 1973, p. 28) The same is true to the whole sphere of law in continental Europe. The Common law of the Anglo-American legal sphere is even less characterized by an axiomatic-deductive structure since it is primarily based on pragmatic case studies, although there are necessarily tendencies towards a deductive systematisation. (About the relation between German, respectively continental law on the one hand and Anglo-American law on the other see e.g., Radbruch, 1946, Goodhart, 1934, Schulz, 1934).

A comparative study of legal methods, as elaborated by Fikentscher, shows that phases of systematizing law do appear everywhere but are less important in the Anglo-American tradition. An extreme example is Holmes who in reference to Ihering's purpose-oriented methodological thinking denied the existence of normative thinking and generally the logical structure of the legal system: Not logics has been the history of common law but rather the experience (See Fikentscher, 1975-77, vol. IV, p. 698).

Those who are denying systematic normative thinking in the Anglo-American world are, however, not more convincing than those who – in continental Europe – refer to norms in an exaggerating way (e.g. Kelsen) (See Fikentscher, 1975-77, vol. II, p. 448).

If we take law realistically, as an "open, historically developing system", then it becomes clear that an axiomatic-deductive representation of law in a strict sense is impossible. The codifications - as stated by the legislator or the authors of codes - are not logically structured systems according to the principle of non-contradictoriness. Therefore, the advocates of concept jurisprudence or juridical logic in today's sense have assigned the logical structuring of legal systems to the legal scientist. Thus Ihering claimed that the principle of non-contradictoriness did not refer to the contradictions produced by legislation, but to contradictions of legal science to itself. According to Kelsen the system of norms of positive law is tested by the logical context of legal statements that are described by legal norms. And Klug stated that it was the task of legal science to axiomatise the systems of the effective positive law and thus to bring them into the structure of exact juristic calculations.

No one would object to this task of legal science (juristic construction and building systems) called since Ihering "higher dynamics" (See Engisch, 1977, p. 5). In all sciences this task serves the purpose of logically unambiguous foundation of knowledge. However, in legal science it has some limitations – as in many other scientific disciplines. These limitations are determined by the complexity of the objects. Human behaviour is the subject matter of legal science as far as it is regulated by juridical norms. The ideal would be a code including all – and only – facts that require regulation in a normative manner allowing to find them easily. Such an ideal case would clearly show which action is permitted and what is to expect as a result of particular action. Such a comprehensive recipe for all admissible actions including all imaginable reactions is utopian. Human action, reaction to an action, and reaction to an reaction cannot be comprised in a survey because their variety couldn't be really described. What is more nobody could foresee the future types of behaviour. Note that even playing chess – a game with clearly defined rules within limited space and including comparatively few actors – allows for an unlimited number of variations. How could one then expect a quantification of human behaviour in general considering that the legal system basically tolerates the free "self-development" of the individual? (Raisch, 1973, p. 27).

No wonder that even optimistic advocates of the idea of axiomatisation (like Klug) can speak only of "quasi-axiomatisation" (Klug, 1966, pp. 149-150) of positive legal systems. A more precise analysis of the "elements" of the legal system makes apparent the character of systematisation in legal science. These elements or parts of law which can be built together into a system in a constructive or a reconstructive manner are the legal norms and concepts.

1.5.1.1. *Logical Structure of the Legal Sentence*

As a whole any particular legal system presented in codes is a system ruling human behaviour. The regulations always refer to a certain matter – particular areas of legal reality, which remain connected with regard to their regulations. Thus, for example, civil law and commercial law are parts of private law. Again these parts can be subdivided into tenant law, family law,

law of succession, etc., expressing the particular rules in legal norms. In this sense, the legal norm is always a part of a system of rules that a priori constitutes a unified system (even if this system is not an axiomatic one). Not one corpus iuris, not one collection of laws is just a "mass". They rather build a system of principles, which is immanently present already on the level of the "mass" of legal norms (See Esser, 1956, p. 227).

It has always been the task of systematizing legal science to present explicitly the inner unity of a positive legal system. Engisch remarked: "What the legal system implicitly includes is explicitly developed by legal knowledge" (See Larenz, 1979, p. 156).

Therefore legislation is not necessarily founded on legal grounds. However this can be demanded by the systematizing, constructing legal scientist (e.g., 19th century scholars studying Pandects or today's authors of juristic text books) as well as by the practical jurist who has to make a decision not only according to the particular case but also according to the system. Nowadays one would hardly claim that the legal system represents an axiomatic-deductive system of connected statements free of contradiction thus perceiving the application of law as a logical process of subsumption. Nevertheless the analysis of the logical structure of legal systems and included norms is still a non-eliminative part of juridical methodology.

Under legal sentence in the widest sense we understand each norm, which regulates human behaviour in the legal system (also including e.g. norms of common law). In a narrower sense the legal sentence is a constitutional legal sentence or legal norm, which is laid down in a particular code. (Kelsen differentiated between legal sentence and legal norm. Thus legal sentences are only the sentences of legal science. The present text doesn't consider this differentiation.) Each legal statement consists of two parts: The empirical fact and the legal result that is regulated by the norm itself.

According to its logical structure the sentence includes two elements, which are linked together by an "if – then relation".

Because of its formal structure in traditional logic such a sentence is called hypothetical judgement, which differs from the categorical judgement

However, the legal sentence is not a conjecture concerning reality or being, but an "ought sentence" or "imperative", (Engisch, 1977, p. 22) so that one speaks of "hypothetical 'ought' sentences". (Engisch, 1977, p. 21) Legal sentences express only an 'ought' limited by the facts.

- Generally there is unanimity in contemporary literature on juridical methodology regarding the purely logical structure of the legal proposition as a hypothetical sentence. Yet there are some discrepancies in legal theory and philosophy. First, the view of the legal statement as an imperative can be interpreted in the sense of a radical "imperative theory"

– imperatives are given by a sovereign authority, which regulates each problem in advance. Because of its normative content referring to the addressee of the norm, one can call the legal statement also an imperative, if one is aware that primarily it is not a historical act of commands, but rather the whole context of single norms which are not arbitrarily formulated, isolated commands, and build, as previously stated, a system that, in an objective sense, is understandable by itself (Esser, 1972, p. 37). Therefore law as a "determining norm" (= imperative) is unimaginable without law as an "evaluating norm". Again this norm proves to be a problem in legal theory and philosophy. However the character of imperativeness remains preserved in the legal sentence. Evaluations as legal norms can dominate the life of humans in the community only through their possible development into imperatives (See Mezger, 1989, p. 240, Engisch, 1977, p. 27).

- Second, unlike Kant's notion of categorical imperative, the view of law as a hypothetical imperative can be misunderstood, namely that it presumes everybody's legal freedom to act in a particular way if he/she is prepared to bear the legal consequences. Hence, any debtor would be free not to pay his/her debts if he/she by the court. Or it would be legally permitted to kill somebody, if the murderer is willing to spend the rest of his/her life in prison. For this reason, Engisch has replaced the term "hypothetical imperative" by the notion of conditional imperative.

Indeed legal sentences are categorical imperatives in the sense of Kant. The result of this view is the so-called principle of legality: In accordance with the code of criminal procedure the persecutor is obliged to interfere in all cases of criminal action – law has to prevent criminality. However on the other hand legal sentences are hypothetical in a totally different sense: In a particular sense each legal sentence represents a hypothesis because it can be applied only under particular constraints of an action – constraints that are given in the sentence (del Vecchio, 1937, p. 59, Engisch, 1977, p. 32, Zippelius, 1971, p. 32). Prohibition of killing, then, should be actually formulated this way: "Except of self-defence, capital punishment and war, killing people is prohibited" (Engisch, 1977, p. 32). One uses the term of conditional imperative to distinguish between this imperative – which is conditioned by particular empirical facts and is to be regarded as categorical – and Kant's hypothetical imperative. This term helps avoiding a too simple interpretation of the logical "if – then relation" as "cause – effect" relation in the sense of the so-called juristic causality which was connected with the "will of the legislator" (who could "willingly" shape this relation) by Zittelmann (1879, pp. 214-216) and Thur (1914, pp. 5-6).

The logical structure of a legal rule equals a conditional clause. The latter consists of an empirical part (the state of facts), which is expressed in descriptive language (as a product of specific juristic abstraction) and a normative part (legal consequence) expressed in prescriptive language. It has to be said that not all rules in a code are complete legal rules. In fact there are many sentences in a code, which are linguistically complete but incomplete as legal rules, being dependent on complete legal rules. We cannot but agree with Engisch's statement: The essential carriers of the meaning of legal order are the prohibitions and commands that are taken out of a code (Engisch, 1977, p. 23).

1.5.1.2. *Logical Structure of Legal Concept*

From the point of view of formal logic each legal rule consists simply of concepts. As elements of the complete legal rule the state of facts as well as the legal consequence are abstract concepts.

All concepts are abstract concepts, because they do not refer to any particular, single object, but to a group of similar objects. Abstraction is the procedure of concept formation, which consists of bringing forward particular characters considered as essential and omitting other, unimportant features of a given object. The process of abstraction provides the basis for the distinction between extensional and intentional view: The intentional view is directed towards the content of a concept. Hence, the concept is to be understood as a complex of essential characters resulting from methodical abstraction and brought constructively to a unit. The extensional view presupposes the intentional, since it is directed only towards a limited extension of the concept defined by the essential characters. (See Kraft, 1960, p. 92) The concept, then, can be understood as a class of individuals clearly limited by its extension. Practically each concept possesses the function of class-building.

So it is neither useful nor justified to suppose an absolute difference between these two views. Basically, each concept can be understood extensionally as a class of individuals, if it is possible to determine the complex of characters common to the individuals in question. Also, the contradiction between the abstract, general and the concrete concept can be seen solely in the extensional perspective. The level of generality or "high degree of abstraction" of a concept follows from "negative abstraction" by omitting some characters. To add some characters leads to specification or concretisation, to omit characters brings generalization in the sense of extension of the concept. In his methodology Larenz referred to Hegel and spoke of "concrete-general concepts" which is a contradiction in itself (Larenz, 1979, pp. 439-441).

Hegel himself distinguished between abstract generality and totality, and assigned the latter only to the phenomena "life", "human mind" and, finally, "absolute mind" or God (See Oeser, pp. 243-245).

Hegel's doctrine of the concept is neither formal logic not epistemology, but – as a philosophy of mind – a special version of modern metaphysics which is not applicable to juristic methodology. The only examples for "concrete general" concepts in Larenz' work – they are also accepted by Engisch in his criticism – are concepts like "people" or "person" (in the sense of "human person"). The latter is placed as a "complete concept" (see Engisch, 1968, p. 174) at the top of a series of concretisations of the juristic concept "person" which can also mean particular organizations.

A "juristic person" is not a person in any ethical sense but a juristic abstraction. Thus, it follows that what is regarded as essential or unessential characteristic is determined by the respective point of view underlying a particular discipline.

Jurisprudence as legal science and practical activity is devoted to human behaviour as far as the latter is regulated by law. The methodological abstraction in jurisprudence aims at giving the essence of concrete "conditions of living".

For example, the juristic term "animal" is to be understood only with regard to the juristically relevant aspects of keeping and breeding animals. Therefore bacteria are not a case for the jurist. Civil law does not subsume the term "animal" under the – more general – notion of living beings. Essential are only legal questions, e.g.: "Who is the owner of an animal?" Thus, animals have usually been considered as movable objects (See Larenz, 1979, p. 433). The extension of the "freedom" of the legislator in this context can be demonstrated by the definition of "fish" in the Bavarian Law of Fishery from 1908: "Fish are" – Linné would be really shocked! – "fishes, crabs, and other useful aquatic animals, as far as they are not defined by hunting law" (After Zippelius, 1971, p. 52).

Legal concepts are indeed basically abstract concepts, but are not always to be found at one and the same level of abstraction. We have to realize that there are different degrees of generality. They follow from the logical relation between higher and lower degrees of order.

The classical examples for such structures of order are given by the 19th century "jurisprudence of concepts". Contemporary German private law too rests to a significant extent on an order of legal terms according to their level of abstraction.

A considerable part of legal concepts, which are used in legal sentences from the point of view of the abstraction of legal rules for human behaviour, are descriptive, classifying concepts with a high or a lower degree of abstraction. The formal-logical structure of such classifying legal concepts is characterized by the extensional-logical view as "class of individuals". One has to take into account that the name of a particular class is a general notion, which is already given; its content or intentional meaning results from abstraction and is determined by definition.

Juristic methodology uses almost exclusively the so-called classical theory of abstraction or the traditional doctrine of definition. The so-called modern theory of abstraction, however, is not based on the classical principle of similarity between the characters or properties of objects, but rather on the principle of equivalence, i.e. the formal equality which cannot directly refer to objects but to verbally or symbolically represented classes of objects. This view is rooted in mathematics. Mathematical objects, i.e. the natural numbers, are classes of equivalence. The modern theory of abstraction, therefore, is to be regarded as linguistically immanent, i.e. as an operation making "conceptual words" precise. Accordingly, "definition" means each way of representing concepts with help of other concepts as far as the standards of the principle of equivalence or formal identity are kept. Thus, definition is just an operation of replacing signs or symbols from the strict point of view of "non-creativity". If the defined includes more information than the defining, then the intention of definition has gone to the wrong direction. The linguistically immanent theory of abstraction or definition supplies jurisprudence with means of systematisation, especially when it comes to the computerization of legal rules and concepts.

However, it does not replace the classical theory of abstraction, which – as a "natural" abstraction (see the results in evolutionary epistemology e.g., Lorenz, 1973, Riedl, 1980, Oeser, 1987) – is not linguistically immanent, but constituting language: It is directed towards concrete facts that are perceived by experience. Moreover, it refers to the application and the gaining of (new) legal concepts – operations, which are strongly connected with one another and have to be regarded as a unified process. Hence, as already Schulz has stated in his historical analysis of Roman law, each decision in a concrete case is more than the solution of a single problem. It is an abstraction for the concrete case in all its specific details cannot return in future. Several steps are required to formulate this abstraction in words. Each step is characterized by classifying general concepts with an ever-higher level of abstraction, but a decreasing content of information.

Such classifying general concepts are the basic elements of any corpus of law, for they clearly determine the respective sphere of validity. In the sense of the traditional doctrine of definition, this methodological postulate means that all necessary and sufficient conditions are to be clearly stated in order to make an object an element of the given class (See Raisch, 1965, p. 36).

Radbruch followed this traditional formulation of classifying concepts. In his – the very first – juristic response to Hempel and Oppenheim (see Hempel, Oppenheim, 1936) he remarked that legislation operates "intentionally with concepts that mark clear lines in the stream of life". Accordingly, legal science has "in principle a classifying character". This means: Legal science forms – for the sake of legal security – classifying hypotheses of concepts at a comparatively high level of abstraction which, in the sense of logic of classes, represent mere extensional structures of order. Classifying concepts indeed promote legal security, but not "individual justice". Therefore, Radbruch claimed another form of concepts in jurisprudence: "Type concepts" which, in his definition, are "structures with a stable centre and variable borders" (Radbruch, 1938, p. 53). His approach goes back to a conception by Heck.

Radbruch's hint at "individual justice" and the historical fact that the metaphors "centre of concept" and "field of concept" are rooted in jurisprudence of interest make clear that the controversy "class concept – type concept" ("order concept") concerns the practical problem of application of legal concepts and is not restricted to formal-logical aspects. Hempel and Oppenheim took physical anthropology and psychology as examples to offer evidence that type concepts are not to be understood as "degenerate", ambiguous concept of classes, but simply belong to another category of concepts; that their goal is not an "intent boundary" in the sense of extensional logic. Type concepts are generally to be seen as comparative concepts, i.e. concepts, which render possible a comparative statement: "more" or "less". They rather aim at establishing series, i.e. systems of order in a particular field, which includes objects with "more" or "less" specific features. Chemistry or meteorology, for instance, deal with objects that require concepts to characterize *processes* and not fixed objects. ("High" and "low" in meteorology, "periods" in chemistry, etc.) (See Oeser, 1978, pp. 34-36). In these disciplines the comparative concept formation – however exact – serves the purpose of a systematisation of given objects and processes.

However in jurisprudence the purpose of typological concept formation (which logically can be subsumed under comparative concept formation) has a primarily practical goal: It has to "leave open" legal concepts in the application of law. There are no definite boundaries, so that the logical structure of "indefinite legal concepts" can be represented.

The logical structure of legal concepts is strongly connected with a legal-theoretical attribute; the same is true to the logical structure of the legal system and the legal rule.

Since law itself is an open system – oriented towards times and the real social life of humans – legal concepts can never be regarded as unequivocal, external classes and concepts. They are rather temporary or temporarily stabilized concepts. Their boundaries are variable and change according to social reality. Each historical state of a legal system and each particular codification include concepts that differ one from another with respect to their specification. The so-called indefinite legal concepts, however, are not indefinable and blurred, but any strict definition and limitation would fail to represent the diversity and complexity of real life to which such concepts have to be applied.

1.5.2. *Application of Law*

Teachers of jurisprudence when speaking of rights and claims distinguish in a case between the question of right (*quid iuris*) and the question of fact (*quid facti*), and while they demand proof of both, they give the proof of the former, which goes to establish right or claim in law, the name *deduction*. (Kant, 1781) Kant proclaimed this distinction to be a kind of methodological manifest for all spheres of rational knowledge and decision. It is still inherent in the structure of each applied law:

The practical jurist is a person who, on one side, treats legal rule, on the other deals with "life" (See Engisch, 1977, p. 43).

Methodologically, this means that each act of applying law has to be "based" on a formal law i.e. justified by legal rules, which are gained through logical inference. Today, this part of the application of law is called foundation of law. With respect to its logical structure foundation of law is a "syllogism of determination of legal consequences" and in its most simple form – a deductive conclusion according to the classical formulation of the "legal question" (*quaestio iuris*) answered by means of deduction. The "question of fact" (*quaestio facti*) has to be distinguished from the legal question. In today's terminology, however, it does not coincide with gaining law. The older terminology was based on the possibility of a separation of the legal question from the question of fact. This was the practical jurist's insurance against all arguments that could make a decision difficult (See Larenz, 1979, pp. 295-296).

Methodologically, "finding law" disconnects the question of fact and the legal question only in very rare cases. The description of a fact already calls for words, which include a legal evaluation like "disturbance of the peace" that is not a mere physical concept. Or "prohibited area" (after Engisch, 1977, pp. 48-49, p. 55) which is not determined by geometric or architectonical principles but means an area which can be walked on only by people having a special permission. Hence, finding law refers to the individual, concrete facts of life but it does so in a specific manner: It stresses the legally relevant points of view.

1.5.2.1. *Argumentation of Law.*

If we take the "model of pure application of law", namely criminal law, the foundation of legal norms seems to follow a rather simple process of deduction (after Engisch, 1977, pp. 45-47), namely:

- Major premise: "If somebody is a murderer, then he/she should be sentenced lifelong to prison."
- Minor premise: "*M* is a murderer."
- Conclusion: "*M* should be put to the prison for the rest of his/her life."

More precisely what is given here is a hypothetical or conditional conclusion in the *modus ponens*: If *p*, then *q*; *p*, therefore *q*.

The major premise is the legal sentence, which is already characterized as a conditional imperative. The minor premise is a descriptive, concrete assumption. Finally, the conclusion is a statement, which is assigned to the sphere of ought. However its realization (in the execution) however is presupposed to be non-problematic. This example demonstrates the basic patterns of the juridical conclusion within the traditional and classical, ie. verbal formulation.

Moreover Klug has shown that the validity of this conclusion is not simply to be accepted. Rather – by the means of calculation – it can be proved and reconstructed in its particular steps, similar to juristic argumentation like "analogy" or *argumentum ad absurdum* that have always been used in juristic methodology (See Klug, 1966, pp. 51-53, 101-103). The executor claims a compulsory foundation. Here, classical logic is sufficient if one does not advocate the Utopia of a computerized process of legal application. Problems do not follow from a lack of formal specification, but from the fact that it is rarely possible to take the super-statement directly from the law and to gain the sub-statement accurately as a descriptive statement concerning an individual case. Between the law and the concrete case there is a chain of major premises and minor premises, which make obvious that founding law does not consist of the trivial construction of a very simple model of conclusion, but rather of the "subsumption" which – as a concept-logical construction – only prepares the conclusion. Thus, a clear distinction must be made between "subsumption" and "subsumptive conclusion":

- Subsumption puts a concept of small extent under a large-extent concept.
- Subsumptive conclusion comes from the possibility to subsume the sub-statement (which contains a concept of small extension) under the concept of large extension of the major premise.

Because of its simple structure (*modus Barbara*) the juristic subsumptive conclusion does not make any problems. The juristic subsumption, however, is the most difficult operation in the foundation of law. This is due to the complex logical structure of legal concepts, which, as was already stated, are only in very rare cases classifying concepts. It is not a priori clear, whether or not a concept of the sub-statement can be subsumed under the concept of a super-statement, so that a specifically juristic *interpretation* is necessary, which is not a formal-logical question and has to be regarded as a problem of finding law, i.e. finding legally relevant aspects of the respective individual fact.

In sum, the logical structure of applying law can be characterized as follows:

- Interpretation is logically pre-supposed by subsumption. Subsumption is the logical precondition to the subsumptive conclusion.
- Generally, interpretation itself has to clarify the "content and extension of legal concepts".
- The content of a legal concept is determined by the definition of (juristically) essential features.
- The extension of a legal concept results from an enumeration of cases, which can be subsumed under the concept in question.

This formal-logical skeleton is sufficiently supported by the instruments of traditional logic (Engisch, 1963, p. 120), even if more complex contexts additionally require a chain of supplementary syllogisms. It is an absolutely necessary part of the application of law, which for reasons of legal security resorts to compulsory foundation. Hence, one whould give right to Forsthoff's radical remark that jurisprudence would destroy itself if it doesn't accept that "legal interpretation is the search for the right subsumption in the sense of a syllogistic conclusion" (1959, p. 41).

The adherence to the logical *Begründungszwang* (coercion for foundation) in the application of law does not mean at all that

- The foundation of law has to be done explicitly by means of syllogism or even calculation,
- It is a linear process, which brings with certainty a solution.

In principle such speculations fail because of the uncertainty of all legal concepts based upon the complexity and uniqueness of all concrete individual facts of life.

1.5.2.2. Finding of Law

The concept of interpretation (*interpretatio*) – as a precondition to the subsumptive conclusion of the formal foundation of law – takes up the traditional juristic methodology as structured by Savigny. He understood "interpretation" in a narrower sense as "legal interpretation". Savigny's gradualist model of interpretation is still in use though in a somehow modified version. The modification concerns the "teleological interpretation" (See Larenz, 1979, pp. 298-300, Canaris, 1969). Together with the grammatical-logical interpretation (referred to by jurisprudence of concepts in a somewhat one-sided manner), the teleological interpretation includes the historical purpose pursued by the legislator. Also it includes the objective purpose (nature and thing) thus accounting for the basic idea of interest and evaluation jurisprudence.

In most of the cases historical interpretation is identified with a "subjective" explanation referring to a particular historical subject as legislator. The will of the latter is responsible for the formulation of legal norms and the meaning of legal concepts. This definition of verbal meaning – which can be also determined by a so-called legal definition (remember the mentioned example of legal definition of self-defence) – must not be transgressed in the sense of a "narrower interpretation" in the search for the will of the historical legislator. Thus by interpretation one cannot subsume a female under the concept of a male person or a gorilla under the legal concept of human being. Nevertheless the legislator – since he/she is free to define according to his/her own requirements – may, as the above mentioned Bavarian Law of Fishery shows, contradict biological classification and extend the legal concept "fish" to "crabs" (See Engisch, 1977, p. 249).

Subjective-historical interpretation has been the object of a lot of criticism. It has been said to deny the systematic and objective task of jurisprudence and to renounce the work of interpreting although this "retreat from responsibility" ends with the fiction of a "reasonable legislator" (See Esser, 1972, p. 129).

Through this objectifying fiction the traditional doctrine of interpretation can transgress the "narrower sphere" of the legislator's mind. It can be an "extensive" as well as a "restrictive" interpretation, as long as it does not contradict law in its literal sense.

The limits of the extensive interpretation are the starting point of analogy. A good example is offered by classical antiquity. It refers to damages that can be caused by an African ostrich. According to Roman Twelve Tables law the owner of a quadruped animal (*quadrupes*) is responsible for damages that might be caused by such a living being. Now neither extensive nor restrictive interpretation can transform a bipedal animal e.g. an ostrich into a quadruped. However, by analogy one can state that what is true to quadrupeds also is also true to bipedal animals as long as they are as dangerous as the four-footed.

Thus analogy does not mean only interpretation and finding law – in the sense of discovering legally relevant aspects of a particular case regulated by the code. Rather it means, "gaining law" in the sense of updating law and requires its own justification (see below). The traditional theory of interpretation takes resort to the basic idea that an "algorithm" – a precise prescription of the series of mental operations (see Kriele, 1967, p. 85) – has to be found in order to come to the concrete result. This basic idea is still alive as Kriele has shown on the example of the discussion about the "law of interpretation" (see Silving, 1958, p. 499, Kriele, 1967, p. 88, p. 91) in the USA. Such a law of interpretation would stabilize the method of interpretation in a way, that all competing methods would become improper; a deviation from this method, then, would always give reason for revision.

For now, it should be obvious that the primary problem of "legal interpretation" is the interpretation of the legal concepts that are used by law as precondition to each subsumptive conclusion. If, and only if, the "interpretation" of legal concepts according to their content and extension is uncontroversial, a clear solution of the respective individual problem is possible. However, an uncontroversial or undisputed interpretation is not an "interpretation" at all. Interpretation is needed only in case of doubts. The different possibilities of applying law are rooted in the ambiguity of legal concepts. Their content and extension is not a priori given, but they are developing in the course of application. Also, with regard to finding law – the evaluation of the legally relevant aspects of a concrete case – we have to consider the primary "uncertainty" of the individual case itself which does not a priori show in which legal concept it is included.

"At the beginning was the case." This statement by Schulz (1934, p. 27) is true to the historical development of any legal order e.g. Roman law, and can also be considered as a principle of the genetic priority of individual facts of life vis-à-vis to the codified law. This means: The jurist has to deal first with the individual case to which legal norms are to be applied. The "topic" method is the search for the very legal sentence that could fit a particular "fact of life", which has to be legally evaluated. In the spirit of Viehweg's *Topik und Jurisprudenz* and the mentioned work by Esser and Kriele it has been sufficiently stated that, unlike the axiomatic-deductive method (which has the goal "to develop, by means of deductive systematisation, jurisprudence to a legal science"), topics is a "theory of juristic practice" (Viehweg, 1963, p. 2). It starts, *ex definitione*, with concrete problems, but pre-supposes the corpus of norms of positive law, where the concrete problems find their "place" (*topos*). The advocates of topics, however, do not agree on the question, how legal norms build a system. Yet they take it for granted that the axiomatic-deductive systematisation in the strict sense completeness does not represent any real possibility. At the same time a general opposition of the "topical view of problems" and "system thinking" is historically as well as systematically unwarranted. Let us consider the three main sources of topics.

They are: Aristotle, *Topika*; Cicero, *De Inventione*; Vico, *De nostri temporis studiorum ratione* (1708). In all these works the original and prevailing meaning of *topos* is the following:

- The *place* in the incomplete system. The notion of system refers to all possible spheres of objects, like law, physics, politics, etc.
- The "pre-fabricated view of argumentation." This mainly rhetorical meaning – developing subsequently – brought *topos* (in the modern age) in opposition to the systems view and led to the refutation of topics as a method "to quibble and talk with fluency and an appearance of profundity" (Kant 1968, pp. 268-269, see also Kriele, 1967, p. 115-116).

Generally, topical thinking is opposed to the view of "closed", i.e. complete and definite systems. The view pursued by Esser and Kriele according to which topical thinking expresses "system openness" of law, is still undisputed. It implies the "permanent correction, extension, and modification" of legal systems. But how can the deductive-logical foundation of law (as syllogism determining legal consequences) be combined with topical thinking (which is oriented towards problems)? There is no clear answer to this question. The refutation of "juristic syllogism" – which, as was already shown, usually displays the simple structure of *modus Barbara* – as a platitude (Kriele here referred to Hegel's "golden words" against Aristotelian syllogistic), is not at all justified (Kriele, 1967, p. 51).

The connection of legal practice to law is necessarily in accordance with the compulsory foundation that in its logical structure represents a deductive system. As already Engisch demonstrated by many concrete examples, deductive inference almost never restricts itself to the unique process of deducting the special legal case from a norm inherent in the law, but rather consists of a number of syllogisms limited by the complexity of the respective case. The gaining of premises as hypotheses of problem solving – which, as super-statements refer to law as sub-statements to the case – makes use of the very "inventory" achievement of topics which, hereby, is always linked with the partial deductive solution procedures. This methodological "folding", which appears in each application of law, was characterized by

Engisch as a "continuously changing perspective" (1963. pp. 14-15) - a metaphor introduced "in a purely logical context in order to avoid a logical circle" (See Engisch, 1977, pp. 210-212). The circle could be found where one is looking for super-statements from the point of view of the concrete case and at the same time considers only those aspects of the case as essential that refer to the super-statement. This procedure is no logical circle in the sense of a *circulus vitiosus*, but rather a feedback process, a flux in both directions: from the concrete case to the system of legal sentences and back again.

The connection between induction and deduction leads to a cycle that does not return without any result to itself, but comes gradually, through step-by-step correction and limitation of possibilities of solution, to a final result. Such a result is given, if the applicability of the considered legal norms to an individual fact of life is guaranteed. But the result is never an absolute end. It is to be understood pragmatically: The process of finding law stops, where a particular action is finally regarded as right or wrong.

The relative uncertainty of legal concepts as well as of the individual, historically unique case justifies the pragmatic breaking-off of the process of decision, which includes justice as a hypothetical goal or task.

The many possible solutions (of a legal case) through the process of finding law can be best shown by the fact that a concrete fact of life becomes a legal case if (and only if) it requires a legal regulation.

The different roles played by claimant and accused, prosecutor and defence attorney, etc. in the legal proceedings clearly demonstrate that there is no a priori solution of a single case, but that a solution *has to be found*: The "subjective right" of the disputing parties has to be presented and defended in such a way that, if possible, the court does not miss any argument or law. Therefore, each struggle for subjective right contributes, as was already pointed out by Ihering, to the development of objective law (See Kriele, 1967, p. 199). The first hypothesis is formulated by the person who brings a case to the court: This person declares some events or facts as legally relevant. However, the first step could also be made by a person who is acquainted with particular legal norms thus being able to classify a case as belonging to a given norm. In this case, no more hypotheses are required. Usually, however, legal cases are not evident violations of law, but follow from doubtful problems that can be solved only gradually in the sense of reducing the complexity of the given individual case.

1.5.3. *Gaining of Law as Updating Law*

Up to now, with the concept of application of law I have presented only those methods that work under the pre-condition of an existing legal system in the sense of codified law. Founding and finding law are based on the assumption that in principle a complete system of norms indeed exists (even if it is not explicitly formulated in details), and that the special case, which needs to be legally regulated can be subsumed under this system. Hence, for each legal case norms can be found within the respective field and its special rules. Codified law includes a "power of order" which makes possible the reduction of the complexity of legal reality by absorbing the individual specificity of the concrete case. However, because of the uncertainty of all legal concepts, the judge has some space to move, so to speak, and to find his/her decision by means of the methods of interpretation. His/her "free space" intended by the legislator to produce "open" legal concepts should not be identified with loopholes in law, which, according to the traditional terminology, means a "lack" (Engisch) or at least an "incompleteness" (Kaufmann) of law requiring completion.

The concept of the loophole is connected with the traditional view of a basically complete, at least intended legal system. Because of the claim of completeness, "loopholes" exist wherever one should have expected a solution of the problems.

The "loophole" is therefore a result of an understanding of law as such; this understanding follows the view of a static system: "In a wall or a fence a hole is given, where its stones or planks stop to exist." (Canaris, 1964, p. 19) There have been objections against this concept of a loophole. It has been argued that this concept is "confusing" rather than "clarifying", because it is tied to the polemic situation of the dogma of the completeness of codification and its upsetting by jurisprudence of interest. Thus Heck claimed that most of the ambiguous legal questions were to be referred to loopholes in law (See Kriele, 1967, p. 196). In what follows, I deal with the methods of completion, extension, and correction of the legal system, and I do not use the concept of the loophole and **its** filling, since, as a static concept, it is incompatible with the legal-theoretical conception of the "open system".

In the open system gaining and updating law – as already Esser showed – can follow only from the "rationality of the system" and the position of the legal applicator in a decision process. Hence, the system will continue to work, when it is filled with all those evaluating criteria and thoughts of order, which appear with all new situations of conflict and expectations. This model of the open system (see the illustration on p. 57) is now essential in our discussion of the methods of gaining and updating law, i. e., methods like

- Juristic analogy,

- Juristic induction, and
- Juristic construction.

The originally – within "jurisprudence of interest" (Heck) – used term *gaining law* (see Heck, 1912) is unlike "finding law" heuristic in character, even in those cases, when it is not used in the sense of updating law. According to the modern theory of finding law (Kriele) and applicative hermeneutics (Esser) each legal application is heuristically constructive for each "explicit order" too depends on the question of understanding and, thus, on the present evaluation program which certainly cannot be simply taken from law (See Esser, 1972, p. 182).

1.5.3.1. *Juristic Analogy*

The traditional theory of interpretation is oriented towards codified law as a static, "closed" system. According to it the starting point of analogy is to be located at the end of extensive interpretation:

It [the extensive interpretation] makes only the thought of law explicit – in opposition to the too narrow expression. However analogy means a further development of the thought, updating law in the direction given by the legal system (Eneccerus-Nipperdey, 1959, p. 340).

In the context of analogy, juristic methodology also includes a discussion of the so-called reversed conclusion (*argumentum e contrario*), because the usually defended view is that, whenever analogy is requested, the *argumentum e contrario* is also at work. Due to similar conditions analogy infers the occurrence of legal consequences. On the basis of different conditions the reversed conclusion infers the non-occurrence of such consequences.

A general legal-theoretical application of the reversed conclusion is Kelsen's "negative norm": I am free to do or not to do, what I am not obliged to do.

1.5.3.2. *Juristic Induction*

In juristic methodology the context of analogy and induction manifests itself in the relation between analogy of law and legal analogy:

- In the case of analogy of law one starts with a particular legal sentence and extends the area of cases, to which it is applied by adding other similar cases.
- Legal analogy is the development of general principles on the basis of several legal norms by means of inductive inference.

According to classical logic analogy means a conclusion from the particular to the particular, so that it always remains at the same logical level. Induction, however, leads from specific to general and is a kind of type leap. Such a leap is not a mere generalization but a creative act, which is characterized by the introduction of a new legal concept. Respective examples can be found in the whole history of Roman law.

1.5.3.3. *Juristic Construction*

By construction in its widest sense one has to understand a rationally established and therefore understandable arrangement of knowledge.

Initially, in its narrower sense, the meaning of construction was defined in geometry (and restricted to operations carried our with rulers and dividers). It is the proof of the existence of a geometric figure by its drawing. Later, construction was extended to arithmetic as a construction of series of symbols. In grammar, the notion of construction refers to syntactic and has been known ever since Cicero (*constructio verbum*).

This ambiguity of construction and its application to different disciplines explains its problem as a method. Its basic preconditions could not provide an absolute criterion for "constructiveness" without the cancellation of the very method. The establishment of such criteria would be identical with axiomatisation. However without these criteria construction remains incomplete with respect to its foundation. But the positive aspect is that the constructive method is open to the invention of new operations of arguing. This openness of constructive strategies is particularly relevant to scientific theories, which with regard to their content do not fit any complete axiomatisation.

Irrespective of its different meanings, construction is a synthetic operation, which pre-supposes the recognition of the elements of a system, but not the structure of the whole, i. e. the arrangement of the elements.

The problem of the juristic construction is the impossibility of a complete axiomatisation of legal order. Hence, founding law is a construction, for it has always to make evident that particular empirical concepts – that refer to legally relevant facts of life – can be subsumed under legal concepts (which can be found in law). As I stated above, this operation only in very rare cases consists of a single subsumptive conclusion based on two clear premises; it is rather a heuristic, creative process which – based on the principle of "unity of legal order" – fits together "the fragments of legal thinking to a mosaic of an explicit super-statement". It is the object relations, which provide arguments for this interdependence:

- One clause should make another clause explicit,
- One clause should complete the others,

- One clause should make its way through another on the basis of exceptions etc. (See Engisch, 1977, p. 68).

With the extent and sophistication of legislation the requirements of construction get stronger. So it is not an exaggeration when Stammler maintains: "Once one has applied a single clause of a code, one has already applied the whole code" (Engisch, 1977, pp. 24-25).

1.6. *Summary: The Open System of Law*

In this summary I try to give a synthesis of the so-called traditional methodology (which is based on the theory of subsumption and interpretation) and the new methodology (which rests on the so-called modern hermeneutics). The following illustrations present the differences between these methodologies as mere changes of the main focus.

To begin with, the so-called traditional methodology, which is still used in German-speaking countries, leads to the following picture:

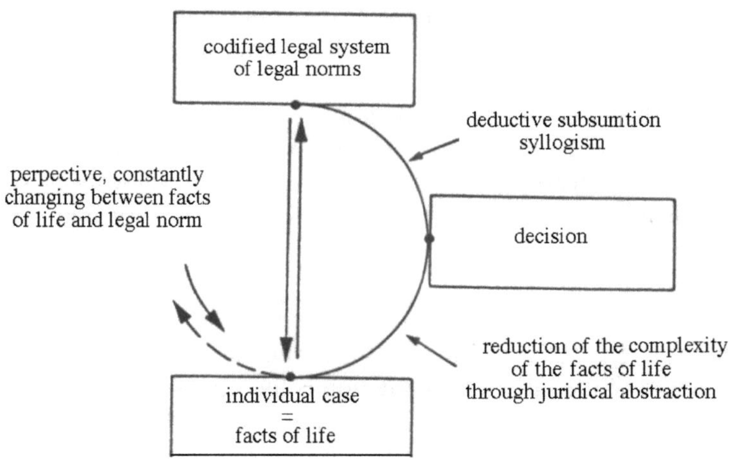

This illustration already considers that the so-called theory of subsumption and interpretation is a correction of the after all untenable model of a pure jurisprudence of construction or concept, presented by students of Roman law (Puchta, Ihering, Windscheid). In this model, which was based upon the fundamental validity of Roman law, the individual facts of life played only a minor role:

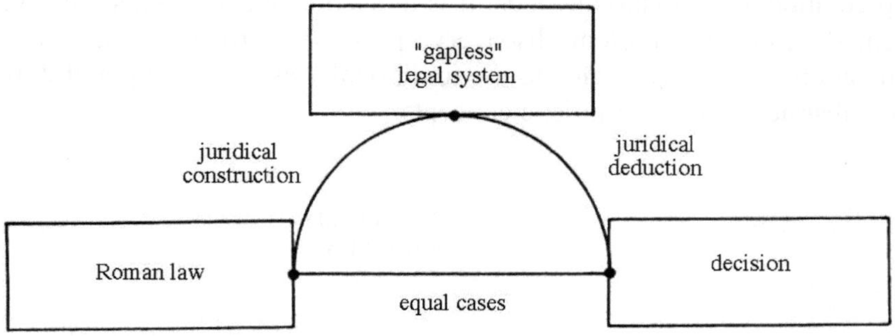

The break down of the ideal of a closed legal system eliminated at the same time the picture of a judge as a "subsumption machine" and made the incompleteness of law evident: The model of jurisprudence of interest, at least in its radical version as the "free-right school" and the Anglo-American case law, strictly opposed the methodology of jurisprudence of concepts and construction. According to this model the systems idea does not play any role. In its extreme version the model even denied general norms and thus returned to the prototype of Roman case law in the sense of "reasoning from case to case":

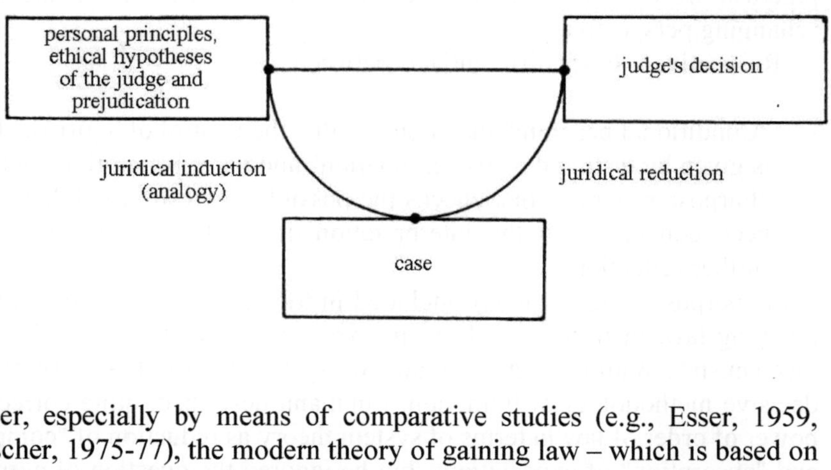

However, especially by means of comparative studies (e.g., Esser, 1959, Fikentscher, 1975-77), the modern theory of gaining law – which is based on the applicative-hermeneutical circle – recognizes that there is actually no difference between the two types of juristic method, namely the "deductive" method of Continental European law and the "inductive" method of Anglo-American case law. After all, both include a mixture of deductive and inductive reasoning (Kriele 1967, p. 200). This insight originates the complete model of juristic methods. It is in accordance to Esser's intuitively presented circle of problem discovery, principle formation, and system stabilization. Also, it is in tune to the traditional view of the "up-and-down" in the "abstract" system of juristic concepts:

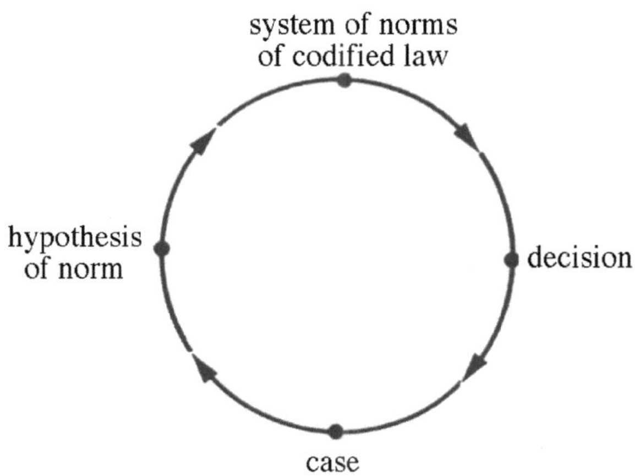

This feedback cycle can be regarded as a correction of Engisch's idea at the level of philosophy of science. It emphasizes the creative elements in the practical work of a judge. This aspect is not yet included in the view of "changing perspective".

Respectively Esser differentiates between

- "Conditional program" that assumes that the control of a juristic decision is given by testing the "if-then relation" and the subsumption, and
- "Purpose program" that checks the possible interests which have not yet been considered in the interpretation of conditions and now call for further reflection.

If the purpose program is not included in the feedback cycle of setting and applying law, then the model of this cycle is reduced to a mere "functional mechanism". With respect to Luhmann's system theory, Esser criticized this decisive methodological limitation. Luhmann described, quite correctly, the power of order of law in terms of system theory as reduction of "complexity" and "absorption" of expectations, but he ignored the question of purpose: In

such conditional programs purpose has no essential function. It is rather the program of different style, for it makes expected effects and not causes constant. This static functionalism has been criticized by Esser from the viewpoint of hermeneutics (Esser, 1972, p. 145).
The misunderstanding of Luhmann's "systems analysis of the legal process" is based on the traditional overestimation of the division of labour between the legislator and the judge. As Esser remarked, this does not fit Luhmann's other theses concerning the universality and interdependence of system formation up to the level of society. If one denies the hermeneutical problems of the functional circle of the juristic method, then Luhmann's system-theoretic conception falls back into the traditional state: The work of the judge is restricted to interpretation of law. The judge takes the judgement from the code, which determines the meaning of the facts relevant to decision. In his decision the judge doesn't any longer follow the notions of true and right that have anyway found their position in a transparent order of social life (See Esser, 1972, p. 213).

Like Luhmann's system theory, the today's favoured technomorphic models of a juristic cybernetics miss the hermeneutical problems, for they grant the juristic system only a neutralizing function of regulating social conflicts. The conflict, then, "does not come to and end by means of a solution, but the solution comes by means of ending" i.e. "the continuation of a conflict is simply prohibited". (See Ballweg, 1970) So it is obvious that a simple metaphor – which seems to reveal "systems isomorphism" – does not make a distinction between jurisprudence and control (in a technical sense) and even does not distinguish between control and self-organization, so that it totally negates the central question of justice. On the contrary, the model of the hermeneutical applicative circle transgresses the program of mere control and ties the decision (concerning a concrete case and following concrete norms) to a "program of evaluation" (Esser). More precisely: The hermeneutical applicative circle aims at overcoming a conflict by means of norms, which fit the given case. Hence, jurisprudence shows itself as a pure self-organizing system; the control (of the steady state) is only a secondary function of the aims. Each solution of a single case improves the legal system. The hermeneutical applicative circle, therefore, is not only a stable feedback principle prohibiting social conflicts, but a developmental process of law, both a growth of efficiency of problem solving and a growth of the legal system itself (in the sense of a non-cumulative evolutionary process). In both cases the hermeneutical applicative circle is evidently not so much a circle, but rather a spiral.

The model of the spiral has been used already by Leibniz and as a "hermeneutical spiral" has found its way to the terminology of juristic methodology. It can be adequately used in two manners of legal development.

With respect to its efficiency in problem solving already Fikentscher using the term of case norm has applied the hermeneutical spiral as a model of the juristic technique of problem solving:

The case norm represents this rule of objective law, which attributes an adequate legal consequence to a fact requiring a solution. The case norm is the legal statement in the technical sense (Fikentscher, 1975-77, p. 202, see also Schroth, 1977, pp. 196-197, Riedl, 1985, 276 – Riedl refers to Schroth in order to establish the connection between evolutionary approach and hermeneutics).

The methodological procedure can be described in the model of the hermeneutical spiral in following words: The formation of the case norm is based on an interruption of the hermeneutical circle at a particular point of the knowledge procedure, between norm and case. This interruption is the result of a continuous "hermeneutical condensation" of norm and case; the relation between norm and case thus becomes closer. With the extension of the legal norm the number of windings of the spiral is increasing. The case norm had gained its definite structure when there is only "one step from the norm to the case and back again", because there is no more open question. The criterion for the case norm, according to Engisch, is therefore the best possible concretisation of the normative with respect to the facts of the given individual case. At this "hermeneutical turning point" no more terminological distinctions are made, because they would not have any meaning for the evaluation of justice. The model of the hermeneutical spiral does not only supply us with a presentation of the methods of legal application in the sense of resolving a case, but also presents the methods of updating law and legal development. A "hermeneutical condensation" leads through a series of normative hypotheses that do not always exist in the code or cannot be constructed on the basis of the given material, but are a matter of creating law. The fundamental achievement of the model of the hermeneutical spiral is that up-dating law does not mean any longer a filling up of loopholes. As already shown, the "traditional" methodology in this context had to transgress the border of its own model. It was not enough to present a continuum between "interpretation" and "filling up loopholes" without any clear distinction, but one had to admit that the "filling up" is nothing but the "continuation of interpretation", however, "at another level" (Larenz, 1979, p. 273, see Esser, 1972, p. 182).

Thus it is also possible without any interpretation to transform the "theory of gaining law" into the "theory of development of law." The "evolutionary" element of the development of law becomes visible if one notices that evaluations can change in time. Hence it follows that the control of correctness can no longer be related to traditional values, but to the present state and predictable future development. The "change of evaluation", then, reveals itself as a methodological phenomenon of legal development. This development is methodologically different from the mere cumulative "supplementation of law" through which the number of legal rules is indeed continuously growing without bringing any fundamental change. Therefore the cumulative growth of supplementation is – if it does not become an information problem – usually always a stabilization of the already existed law. However, the evolutionary progress of law is always connected with a basic change of legal concepts and a new organization of the complete (legal) system. Philosophy of science characterizes the methodological difference between merely cumulative and evolutionary growth in any field of knowledge (See Oeser, 1979, p. 170) by the following terms:

- *Negative feedback* with systems maintaining and systems stabilizing effects because each case can be basically adapted to the system;
- *Positive feedback,* which leads to developmental changes of structures.

Whereas the model of systems stabilization and cumulative extension of law remains at the level of a feedback cycle, the structure of a legal development can be adequately represented only by the model of the spiral.

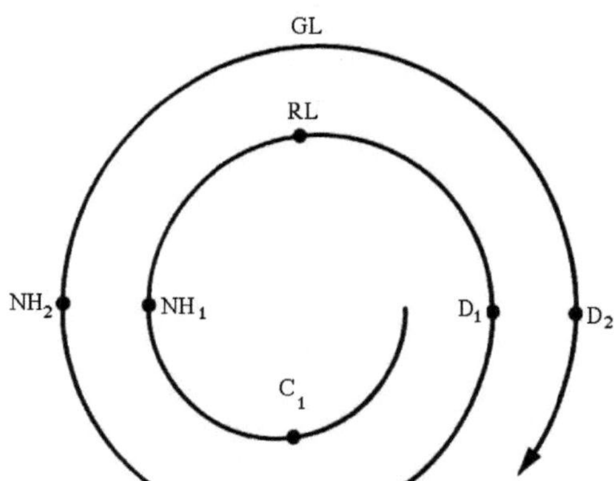

As typical example of such a process is the development of contemporary German law from the Roman law. It clearly shows what was already seen by Savigny: Contemporary German law (GL) is not a continuation of Roman law (RL) although it contains many of its elements (NH), but an autonomous systematisation that "originally" unifies various elements and sources.

2. LEGAL PHILOSOPHY AS A DEVELOPMENTAL THEORY OF LAW

2.1. *Legal-Philosophical Relativism*

Contemplating the whole history of legal philosophy – from the classical theory of natural law to its further development in mediaeval Christianity to the law of reason of modern times – one should expect to find at the end of this longer than 2000 years period of time, the 'victory' of practical reason reflected in the sound existence of fundamental legal principles. Instead, one is rather confronted with the disownment of reason through legal-philosophical relativism.

The first and most comprehensive diagnosis in this aspect is offered by Emge's *Habilitationsarbeit* (On the Basic Dogma of Legal-Philosophical Relativism), published in 1916. The merit of this treatise is that it digs out the common gist of all relativistic formulae. Hence, Emge's view on the basic dogmas of legal-philosophical relativism preserves its validity for all later versions and could be summarized as follows: Being not provable, the value-judgements are not universally valid as well (See Engisch, 1971, p. 247).

According to Engisch, this reflects the "quintessence" of every legal-philosophical relativism, as shown in the statements of its most prominent advocates:

- Max Weber: Empirical science cannot teach anybody what he or she ought to, but merely, what he or she can, and - under particular circumstances, - what he or she wants.
- Gustav Radbruch: The last ought-norms "concern not the knowledge but rather the confession".
- Hermann Kantorowicz: The world of values knows no logical coercion, "nobody could be forced to value an object less or more", the values are always values "to me, to you, to us, to historically given individual or collective personalities".
- Hans Kelsen: "The judgement proclaiming something for just could never claim to rule out the possibility of an opposite judgement".

All the above statements point out not to the real diversity of value systems, but rather (as Emge correctly recognized), to "the dogma of non-provability which constitutes the peculiarity of legal-philosophical relativism" (Emge, 1916, p. 23).

Yet one has to differentiate between the principally dogmatic, legal-philosophical relativism on the one hand and actual relativism, on the other. The latter can be empirically established in legal history on the basis of comparison of law and it reflects the fact that the differences between individuals and practical situations lead to individually different values and demands. The empirical relativism is harmless in comparison to relativism in principle, which maintains that the so-called value judgement is – in logical sense – no judgement at all with a claim on truth and provability. Still this relativism has to be taken seriously, the state of facts being that almost all legal-philosophical relativists refer to the empirically established diversity of past and present legal systems. In other words, the actual empirical diversity of judgements and legal principles constitute the frames for a deeper-reaching dogmatic relativism, which, according to Kriele, represents a theory of basic unreasonableness.

Which are the practical consequences, stimulating the fight against and the refutation of legal-philosophical relativism? These consequences are perhaps best seen in the appeal of Gustav Radbruch, a prominent jurists, who, considering his own position in legal philosophy describes himself as a convinced relativist. Thus, in 1945, in a short treatise *Fünf Minuten Rechtsphilosophie* (Five Minutes of Legal Philosophy), influenced by the non-constitutional state of national socialism, elucidated: If the assassination of a political adversary is honoured, if murder of other races is imperative, and if, at the same time, the same acts accomplished against your own partisans are punished with utmost cruelty and humiliation, then there is neither justice nor law (Quoted after Engisch, 1971, p. 254). However this means that there exist absolutely valid legal principles, stronger than any legal statute, which actually are old acquaintances of ours – namely what one would call natural law or law of reason. Concretely, these principles emerge – analogously formulated – in all declarations of human or civil rights.

No matter how impressive and moving the appeal for human rights may be, one should not be as stated by Engisch (1971, p. 254), satisfied. The refutation of legal-philosophical relativism demands a penetrating and progressive argumentation, which cannot be adequately answered by a merely plausible presentation of its practical consequences.

The first step in this argumentation concerns the actual empirical frames of every dogmatic legal-philosophical relativism. The way one could establish (with the help of a rational reconstruction of legal history and comparative leagl studies), the historical in time and contemporary in space diversity of legal principles and the respective value concepts, so one could elaborate the historical continuity and synchronic convergences of legal systems, despite of the whole discontinuity and divergence. In this context,

Hegel's words – look reasonably at the world, and the world will look reasonably at you (Kriele, 1979, p. 21) are often quoted.

The second step concerns the dogmatic core of relativism, e.g. the claim that value judgements are not provable. The only worthy opponent to this theory of unreasonableness is a legal philosophy as a theory of practical reason. On the other hand, such a theory cannot be itself based on a dogmatic metaphysics in the sense of a new legal ontology or ethics of the material values. The solution meant here is rather a "Kantian" one – a law of reason as a regulative idea. What, however, this regulative idea (of the rightful law) needs is a developmental theory of law as well as argumentation-theoretical concept of justice.

2.1.1. *Empirical Relativism and Its Overcoming by the Comparative Method*

Empirical relativism rests on the fact that law, on the one hand, differs from country to country, and, it undergoes constant changes with the time – on the other. This fact is reflected in the existence of two legal scientific disciplines, which, by embracing each a different approach – a synchronic and a diachronic one – set against one another the corresponding legal systems:

- The diachronic approach is encompassed by legal history.
- The synchronic approach – by comparative law studies.

The diachronic approach, focussing itself on legal systems placed differently in time, grows out of the practical urge to establish the historical limitations of a contemporary legal institution by comparing its past and present structures and functions.

Hence, legal history concentrates itself more on the continuity in the development of the legal system than on its changes.

This is well illustrated by the historical recourse to Roman law in the case of German and Italian jurisprudence. Contrary to the situation in the USA and Great Britain (See Stevenson, 1950, p. 618, Hamson, 1950, p. 672) where Roman law counts as a foreign law, here it has become an element of the national law. It had been adopted by the Italian scholars in the 12th century in Bologna, and by the German jurists (especially by the historical school e.g., Savigny, Puchta and Ihering as well) in the period from the 15th up to the 19th centuries. Thus, bearing in mind that the civil code has been borrowed from the *Pandects*, a lot of the civil legal institutions can be made comprehensive only when compared to Roman law (See Kaser, 1967 pp. 337-339).

The diachronic approach of legal history, limiting itself on national scale, has to be distinguished from comparative law studies, the latter systematically opposing to one another different legal systems and legal circles. The synchronic approach aims not necessarily only to simultaneously existing legal systems. More essential is the requirement that the compared objectives should be found in the same developmental stage (See Pollock, 1903, p. 76). This would comply with one of the general basic postulates of comparative methodology, taken into account not only in the hermeneutical humanities (comparative linguistics, literary studies), and in the natural sciences (comparative anatomy, behavioural sciences, etc.), but also in philosophy of science (See Rheinstein, 1974, pp. 30-31). Ultimate goal of each comparison is the reduction of the compared objects to one basic structure (See Oeser, 1982) This goal brings together both the synchronic and diachronic approaches (respectively legal history and comparative law), which at the end serves the unification of law. That's why the historical origins of comparative law are to be found nowhere else but in the philosophical natural law doctrines of the 17th and 18th centuries.

The French philosopher and mathematician Pascal, disappointed from the historical and national-scale non-unanimity of legal systems, declared critically (long before Kirchmann): There is almost nothing legitimate or illegitimate that would not alter its nature with the quarter of the heavens. Three degrees latitude upset the whole jurisprudence. A meridian determines the truth. Some few years of tenure change fundamental laws. Law has its epochs. A really ludicrous judiciary limited by a river or a mountain (Quoted after Engisch, 1977).

Nowadays, the natural law basic idea about an absolutely, out of time, and trans-nationally valid law can no longer be the fundament of modern legal history and comparative law studies. Despite of all unification tendencies, the aspects in the different legal systems, which as a result of their historical development cannot be unified or harmonized, are clearer than ever (See Wahl, 1950, pp. 298-299). Nevertheless, one cannot but recognize the contribution of historical-comparative methodology to standardization of juristic argumentation and decision-making processes. It is comparative law methodology that makes possible to reveal the system character of law in its functional basic structure. Thus Esser summarized the results of his comparative law study in the following way:

All cultures with legislation show the same pattern of circulation between discovery of problems, postulation of principles, and stabilization of the system. What one perceives – after the widely held methodological dispute – supposedly as a competition between 'open', problematic thinking on the one hand, and 'closed', conceptual-deductive thinking – on the other, is the necessary and in batches 'metabolism' between the new experiences out of the case studies and the forming power of the school (Esser, 1956, p. 7).

From philosophy of science perspective this view is substantiated on the basis of the functional connection between induction and deduction, as referred to by Esser himself (1956, p. 47).

History of law and comparative law do not render the answer to the philosophical question about the just law. Nevertheless, they provide for the empirical foundation to a general legal theory, which doesn't separate the methodological problem about the functioning of positive law from the problem about the evaluating basic norms which determine the adjudication. These, not yet established in positive law, basic norms originate neither from the positively staid law nor from a superiorly placed speculative system of legal philosophy. Rather, they are the result of the interaction between staid norm and legal reality.

This, however, doesn't totally negate the idea of a philosophical legal concept as a measure of any comparison. It means solely that the just or trans-legal law concept of legal philosophy, seen as a theory of practical reason, is not more than a 'regulative idea' in the spirit of Kant. At the end, it is the only way to define *Tertium comparationis* as a basis for comparison between the actually existing legal systems. Main representative of this view is Radbruch (1906, pp. 422-423) who supported the opinion that each comparative law study needs an ideal system as a reference – an opinion, strongly contested by the advocates of empirical comparative law (Constantinesco, 1971, p. 9).

What this controversy reflects is basically the difference between empirical legal science and legal philosophy. Thus, empirical legal science, having as a subject-matter positive law, transforms itself into a comparative legal science by applying comparative methods. As such, its only task is to bring order in the chaotic diversity of legal systems and in the standardization of each particular legal system in the various legal circles (Constantinesco, 1971, p. 273).

Other is the case with legal philosophy, where the application of comparative methods serves a particular goal, namely, it has to confirm the realization of the concept of just law in legal history, the task of the latter being the discovery of developmental rules of law.

The earlier stages of comparative law studies didn't differentiate between those two goals but were dedicated to the search of developmental laws. That search was decisively stimulated by the biological evolutionary theory, which was the first to render a natural-scientific basis to the speculative conceptions of socio-cultural evolution.

2.2. *The Theory of Socio-Cultural Evolution as Frames of a Developmental Theory of Law*

The historical origins of socio-cultural evolutionary theory can be traced back to a period before Darwin. Its initial impetus, as Childe (1975) stated, was provided by the studies of European scholars on savages. The astonishing diversity of social structures, economic organizations, and legal systems in those societies brought the researchers to the idea of different stages of developmental. As early as 1768, in his *Essay on the History of Civil Society*, Ferguson opposed the stages of savages and barbarians to the one of civilization. Before the publication of Darwin's *On the Origin of Species* (1859), Spencer offered in his *Social Static* an analogy between society and organism. Under the influence of Darwin, Spencer elaborated this analogy extensively in his *Principles of Sociology*. Spencer's basic idea is that human societies with their structures 'grow' in the same way as the organisms themselves. According to his view, the still existing savage and barbarian societies had been hindered in their growth, which permits us some glances in the earlier stages of socio-cultural evolution.

Tylor formulated the idea of the socio-cultural evolution even more comprehensively than Spencer. According to his view, social organizations replace each other like "deposit layers in the rock", and at that – uniformly all over the world, no matter of racial and language differences, determined by the same human nature (Quoted after Childe 1975, p. 17).

The American Lewis Henry Morgan's presentation of socio-cultural evolution, although similar to those of the English evolutionists, won, nevertheless a particular significance thanks to Marx and Engels, who borrowed Morgan's scheme. Like Tylor, Morgan was convinced that all three stages – savages, barbarians, and civilization – reflected a forgone but real historical process, which could be reconstructed the help of comparative methodology. He claimed, that the internal organization of our savage and barbarian ancestors was still so well illustrated by some aspects of the contemporary human family that the separate stages of this progress were relatively well preserved (Quoted after Childe 1975, p. 20).

Still, neither Morgan nor any of his forerunners could explain the transition from one stage to another. Friedrich Engels was the first to offer such an explanation. In his book *The Origin of the Family, Private Property, and the State* he developed a theory, linking the socio-cultural evolutionary stages with Marx's concept of the material forces of production. That link, however, was not to be maintained. Right upon Darwin's *On the Origin of Species*, Marx wrote to Engels that the book "renders natural-historical basis for our views". Two years later, however, and after exacter examination, he accused Darwin (in another letter to Engels) of applying social-scientific categories to nature and of presenting "the animal kingdom as a civil society". Actually, Marx's critique was aimed at Darwin's reference to Maltus, who, according also to Engels, belonged ideologically to the "modern civil society". That was the first time, (although perhaps unintentionally) that the question about the historical priority of biological and socio-cultural evolutionary theory arose, a question to be later strongly emphasized by Hayek. He defended the opinion that Darwin borrowed the fundamental concept of evolution from the sphere of "cultural events" like language, law, and morality (known from earlier times through the works of Adam Smith and David Hume), simply applying it to nature. In that, he contested heartily the "secondary" application of this theory in its "naturalized form" to social sciences: "Those social scientists, who in the 19th century needed Darwin to learn what they should have known about their ancestors, nowadays, through their 'social Darwinism', render the progress in the theory of cultural evolution a bad service" (Hayek, 1983, p. 173).

Yet, after a closer look, one can easily see that this priority question has no more sense than the proverbial question about the hen and the egg. Thus Malthus, the very partisan of national economy, (quoted by Darwin himself) thought "biologically" and didn't separate human life in society from the wholeness of nature. Contemplating about the situation in nature and in human society, he discovered the same "law", responsible for the waste in the world of plants and animals and the "poverty and depravity" among humans (Quoted after Oeser, 1974, p. 80).

As Hayek pointed out the same was true to Hume who didn't look upon humans as "an exception in the pool of all existing animals" and who consequently refuted the separation of "natural" from "artificial" (the latter being Hume's expression for cultural).

2.2.1. *Relation Between Biological and Socio-Cultural Theory of Evolution from a Systematic Point of View*

The fact that the relation between biological and socio-cultural evolutionary theory cannot be possibly explained on the ground of a genetic-historical priority, doesn't directly influence the establishment of a theory of socio-cultural evolution. The genetic-historical priority being no logical priority can hardly be the reason for the logical deduction of one theory from another. The systematic relation between socio-cultural and biological evolutionary theory is to be much better understood as an expanding self-development of two parallel to each other fields of knowledge. The similarities between them allow for a heuristically valuable exchange of methods and results. On the other side, such an exchange can prove totally misleading as the left or right oriented social Darwinism, or, for that matter, the seeming biologically sound racism demonstrate. This interdisciplinary formulated problem- circle, originated from the mutual reference of two theories, stemming from different scientific fields, shows the necessity of a model of philosophy of science, 'legalizing' the mutual transfer of methodological and with respect to the contents views. No matter that this necessity has been recognized by a lot of authors, still there exists no generally accepted philosophy of science, but just the dogma, concerning the barrier between sciences and humanities. Obviously, the adequate answer cannot be found in a kind of formalist philosophy of science, which acquires its commonly valid character only when acting as an applied formal logic, and which systematically locks out all heuristic (with regard to the contents) questions, limiting itself to context of justification.

The only acceptable solution, revealing itself through the internal development of philosophy of science, is the aged and forgotten comparative philosophy of science, propagated simultaneously but independently by Oppenheim and Lewin in the twenties. By comparative philosophy of science, both authors understand a systematisation based on the method of comparison, which topologically (e.g. not hierarchically) orders the types of science according to their methodological similarities (See Oeser, 1982). Such a philosophy of science evades the usual approach to the relation between the socio-cultural and biological evolutionary theory. This approach practically consists of a generalization on the basis of either 'upwards' or 'downwards' reduction and inevitably provokes the 'priority' debate. The comparative philosophy of science concentrates instead on revealing the elementary, basic structures of rational cognition and action, the former determining – in various combinations and with different importance – the specific subject-matter- and problem-oriented processes in the particular sciences.

The diachronic as well as the synchronic comparison doesn't therefore cause the dissolving of philosophy of science into a vague, non-committal methodological pluralism. Rather, it reveals the developmental and self-determining mechanisms of each particular science, which represents by itself the fundamental prerequisite for a methodological or with regard to the contents transfer of results from one field to another.

This reduces the antagonism (till nowadays dramatically seen as a gap between two cultures) between natural sciences on the one hand, and humanities, respectively social sciences – on the other, to a surmountable, open to relativism, methodological distinction. Naturally, each transfer or binding between both groups of sciences has to be still accompanied by a methodologically determined process of argumentation and justification to be realized step by step at the different levels of comparison. Stafford Beer (1966) offers a suitable systematisation of these levels of comparison, distinguishing between three grades of similarities (Semmel, 1984, pp. 25-26).

The first level of comparison, although 'superficial' and mostly unstable, is nevertheless heuristically valuable. Here one can speak about a metaphoric similarity (of the weakest grade), focussing on a specific perspective of the comparable sphere of objects, and not testable itself in the reality. Such metaphors have – as already shown – mutually dominated the relation between the biological and socio-cultural theory of evolution.

The second level of comparison – a more precise one – is that of analogy. At this level the congruent characteristics are distinguished from the other features so that one can determine the scope, value and efficiency of an analogy. In this sense there can be (as Konrad Lorenz remarked) no wrong analogy but only a more or less detailed, and therefore, more or less informative one (Lorenz, 1974).

The third level, according to Beer, is the one of isomorphism – the most abstract of all. As such it is not descriptive. On the basis of abstract models of the compared spheres of objects, it rather serves the extrapolation of identical trans-disciplinary regularities relevant to those system properties, which are used in the model construction.

As already shown (Oeser, 1987, pp. 19-21), such ventures have been initially not very encouraging. They were necessarily based on the biological theory of evolution, the latter considered even by those advocating the priority of the socio-cultural evolutionary theory (as F. v. Hayek) to be the concrete and indispensable elaboration of the general idea of evolution. The expansion of the biological theory of evolution 'upwards' to a theory of socio-cultural evolution, and 'downwards' - to a theory of pre-biotic evolution, was seen by Darwin himself as a logical consequence of his own speculations. Thus, he had always hoped, that "one day the principle of life will be recognized as a part or consequence of a general law". In fact Spencer formulated the first version of such an universal evolutionary law at Darwin's time. Shortly paraphrased it reads: The entire perceptible reality constitutes a transition process that leads from an indefinite unconnected homology to a definite connected heterogeneity. (Spencer, 1862) One of the best-known evolutionary biologists of our time, Ernst Mayr, characterized this formulation as "nonsense". Obviously Darwin himself couldn't find much sympathy with it, although he referred frequently to Spencer. He admitted that the expansion of the biological theory of evolution was for him an non-soluble problem to be postponed to a "distant future" claiming that the question about the way the mental capacities of lower organisms developed was not less hopeless than the problem of the origins of life itself.

Yet the above mentioned shows clearly that the tasks Darwin and Spencer formulated differ from one another. Thus, Darwin was interested in concrete solutions while Spencer – in general laws of constraint. These two tasks – each of which is justified for itself, as well as with respect to the other – still represent distinct problems. Yet, they are not, as many wrongly state (e.g. Hayek, Toulmin, etc.), opposites. Thus, neither Darwin denied the existence of general developmental laws, nor Spencer – the principle of selection. Not only philosophers, economists and sociologists like Hegel, Marx, and Comte, differentiated between various stages of development. Biologists, like Haeckel for example, also assumed in the frames of the biological theory of evolution the existence of such stages or phases at the level of ontology as well as at the level of phylogeny. What is more, having in mind the pre-biotic evolution, contemporary evolutionary theory can be no longer limited only to biology. It actually gives an explicit justification of Spencer's universal evolutionary law. Thus, Glansdorff and Prigogine refer to its oldest version (of 1862), reviving it in the following way: Development is integration of matter and an accompanying dissipation of motion (Glansdorff, Prigogine, 1971).

The biological theory of evolution in the 20th century (up to nowadays) has also shown that the assumption of general evolutionary regularities, bringing about demarcation of different stages, is indispensable in characterizing time spans. The older evolutionary theorists differentiate between typogenesis and typostasis. Today, the concept of 'punctuated equilibrium' in the frames of biological macroevolution permits the presupposition of changes through 'leaps' in the evolutionary process. This, consequently, allows for the distinction of phases, reflecting changes in the 'direction' of evolution. Or, to put it briefly, nothing can be said against the assumption of such regularities as long as they are not misunderstood as deterministic laws. Thanks to the acquired knowledge about the statistical character of evolutionary laws and their relation to unique historical marginal conditions, the alternative between absolutely deterministic natural laws of evolution, permitting definite prognosis for the future development on the one hand, and a blind, without a direction process, causing in earlier times a lot of worries – on the other, doesn't exist any more.

To turn back to Spencer and Darwin, one has to say that Spencer's fundamental idea of increasing complexity, although not offering a sufficient explanation, is indispensable in understanding socio-cultural evolution. Thus, Spencer's formula "from incoherent homogeneity to coherent heterogeneity" describes processes of the self-maintenance of a system in its environment, surviving thanks to its increasing complexity. As already pinpointed by Freeman (1974, pp. 211-237), this means that evolution is steered by the activity of self-organizing organisms, and not primarily by natural selection, i.e. by the environment. This approach emphasizes the organism's capacity for learning, which is fundamental for socio-cultural evolution, whereas Darwin with his selection principle asserts a mechanism, operating efficiently "behind the back" (as adequately stated [Eder, 1980]), of the self-developing species. Nevertheless, it would be misleading to oppose Darwin as a pure "selectionist" to Spencer, who with his speculations on the increasing complexity on the basis of stronger coherence between different elements, anticipated modern system theory. Thus, Darwin, at least with regard to socio-cultural evolution, dealt with in his book on the descent of man, goes beyond the principle of selection. He characterizes, and at that even clearer than Spencer, the process of social and moral development of humans as a phenomenon of cooperation in the modern sense of the word. Speculating that no tribe could exist as such if murder, robbery and betrayal were generally spread, Darwin remarked that the struggle for existence didn't occur at the level of the individual but at the level of the group, in the boundaries of which social virtues developed that could then spread themselves to ever bigger communities thus reaching the people from all nations and races (See Oeser 1988, p. 163).

2.2.2. *Segments of Socio-Cultural Evolution*

Already the English evolutionists – from Spencer to Tylor – have been criticized not to consider any particular culture in its totality but, rather, merely isolated components of cultures. Still, a general theory of socio-cultural evolution accounting for all cultures and societies of humankind can be established only on the basis of a concrete comparative analysis of particular segments of human culture, e.g., science, morality, law or economy. At that, the danger to produce a "patched up cultural theory" (Childe 1975) is smaller (and is to be considered as a necessary evil), than the danger to sink in vague and blurred metaphorical generalizations. The task of a theory of socio-cultural evolution would be to trace the development of cultural universals and to found them theoretically. The easiest way to achieve this is to choose a sphere, characterized, on principle, by a certain cultural invariant, i.e. the sphere of scientific knowledge (see Oeser 1987, 1988), which doesn't vary from culture to culture. Hence, quite correctly one speaks of a 'scientific universalism', based on the refutation of ambiguous 'truth' and on the inter-subjectivity of the whole scientifically founded knowledge. Thus, one can (re)construct a 'spiral of learning', which is cross-culturally invariant and the result of an iterating process of cognitive achievements and which, ever since Plato, has manifested itself in an exact differentiation and hierarchism of the forms of knowledge. From Plato originated also the fundament of the natural sciences of modern times, which, especially with regard to their most prominent representatives, Galileo and Kepler started as a 'mathematical Platonism'.

The presented bellow graph shows the following:

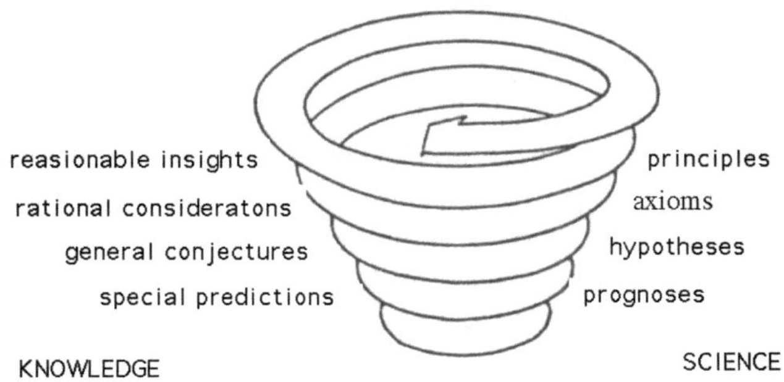

Knowledge, be it scientific or pre-scientific, serves the efficiency of the prognostic which is vital to our existence. Knowledge gaining always means gaining of anticipatory knowledge, which steers our future actions. From special prognoses one can move (for economical reasons) to general conjectures, corroborated by rational considerations and getting almost dogmatic-trivial character on the basis of insights of reason. The same process, however – over many generations, repeats itself in history of science: There one can also prove that the improvement of the prognostic has been the actual impetus behind any systematically regulated observation. It is not really by chance, that at the beginnings of science one finds the pre-theoretical Babylonian astronomy, succeeding without any theoretical foundations but only with the help of extrapolation of a series of observations to offer remarkable prognoses, on the position of the planets as well as on the solar and lunar eclipses. Yet, those were singular prognoses without a general character and referring to particular objects. It was the Greeks to attain generalizations in the sense of scientific hypotheses. Thus, on the example of the Euclidian geometry, they basically anticipated the structure of a scientific theory with its proceeding from axioms hierarchically ordered, deductive mechanisms. On this basis the modern fundamental theories have been elaborated as theories of principles, which in itself is no longer a question of observation and induction.

2.2.3. *Evolution of Morality and Law*

The same is valid also for the second in size sphere of socio-cultural evolution – the one of morality and law. Morality and law have reached their comparative stage of development not on the basis of social contracts – as Rousseau or Hobbes seemed to believe (or at least presented as a symbolical explanation). Rather, they have developed 'evolutionary' in the sense of Hayek's theory of complex systems. Or, the development of morality and law is essentially a "process in which individuals are chosen in the same way by chance and fate as in the biological evolution" (Hayek, 1983, p. 175). The rationalization of these processes on the basis of explicitly reflected rules, laws or norms came later. This can be illustrated with a graph similar to the one of knowledge and science:

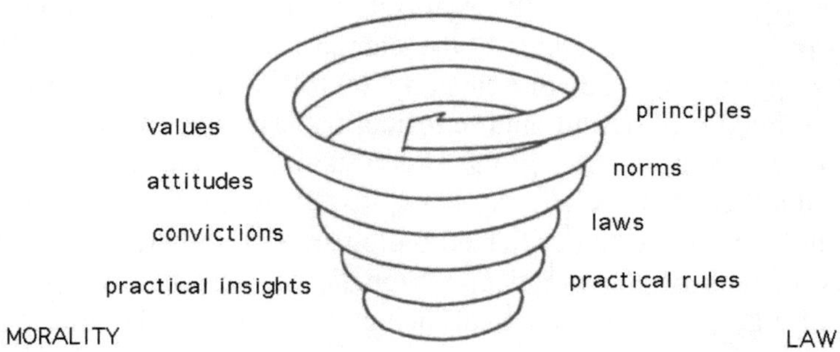

This scheme makes it possible to turn to an – nowadays almost unknown – episode of the German legal history. The idea that the development of law is a process of self-enhancement after the Darwinian model of biological evolution is traced back to the distinguished jurists Ihering. His work *The Purpose in Law* (1877) transformed the German jurisprudence of concepts into a jurisprudence of interests. Some years earlier, in 1872 in Vienna, Ihering gave a public lecture, the title of which – *The Struggle for Law* – reminds one of Darwin's theory. The exceptional success of the lecture – the manuscript was forthwith translated in 18 languages and lived through numerous editions – demonstrates that Ihering's main idea, namely, the evolutionary explanation of the origins and development of law, proved to be of great importance. Contemporary expounders, as for example Wieacker (1981), see in this explanation "the correct assumption that an efficient law should continuously integrate to and up-date itself according to the legal assertions of the individual". That is why 'the struggle for law' is a true selection process, a mechanism, which leads by itself to an improvement of social life, based on the fact that only the patterns of behaviour, that prove to be effective in history can predominate. Hence, according to Ihering, one doesn't need higher and abstract ideals like, for example, humaneness to bind oneself to behaviour rules. Rather, it is the egoistic drive of self-maintenance of the whole society, which with coercive power guarantees the individual her or his rights. Dilthey saw in this position the fall of Ihering in "the murderer's cave of German Darwinism" and commented in a letter to von Wartenburg in 1878 (Wieacker, 1981, p. 349): "...again a clever brain has been killed". Still, one can clearly see that Ihering's view is hardly a trivial metaphor, similar to those offered by the left and right wing social Darwinism. It is much more an explanatory model, providing stable arguments in support of the assumption that the subjective feeling of 'right' is part and parcel of an inter- and trans-subjective system of law, emerged on the basis of a process of self-organization. Taking into account that the starting point of Kant's ethics – the categorical imperative – reflects the generalizing capacity, i.e. the capacity to follow laws and norms, of the subjective maxim of action, and, considering Darwin's explicit reference to Kant (*The Descent of Man*) in the elaboration of a concept of evolutionary ethics and a theory of sociality and law, it is clear that this evolutionary way of arguing is not the often criticized biologism but, rather, an explanation of the very origins of the practical reason. These origins are rooted in the sphere of morality with practical insights which can be traced back to subjective reflections on real actions and which are then to be legally transformed into practical rules to serve as empirically sustained guide-lines of human behaviour. They are no written rules, nevertheless, they are binding for the members of a particular group. The practical insights are superimposed by subjective convictions, forming the moral identity of a person and, at the same time, acting as laws in the legal sphere. However, these laws are *a posteriori* established by special authorities, which makes them purposefully changeable. The subjective convictions originate then inter-subjective attitudes, which at least can be made clear to other people even if these attitudes are not accepted by everybody. Respectively, in the sphere of written law, the generally binding norms claim particular actions or permit or prohibit other actions. It is only after the realization of all these developmental stages, leading to more precise and stable laws and norms that the values and principles come to existence. (A similar systematisation yet without an evolutionary background can be found by Schwemmer, 1986, pp. 83-85). The establishment of Western ethics, i.e. in the sense of Plato's doctrine of ideas, starts at this level. Therefore the very beginning of ethics in the time coincides with the relative end of morality as evolutive process, or, ethics is the reflection of theoretical reason upon the development of practical reason. Likewise, legal philosophy is a theory of practical reason in the sense that it reflects only what has already been established in the sphere of practical jurisprudence, or, more precisely – in the decision-making model of the judge.

The starting point of the evolutionism of legal ethnology differs completely from the one of Ihering's mainly methodological concept.

In 1861 Sir Henry Sumner Maine – considered in England as the founder of legal ethnology and comparative law (See Stein, 1980, Constantinesco, 1972, pp. 149-150) – came forth with a justification of the comparative jurisprudence, aiming at revealing the laws of legal development by using the method of comparison. Being an advocate of the theory of evolution, he tried to trace the beginning of this development in 'early law' (Maine, 1861). The idea of a natural fundament of law was even stronger defended by A. H. Post. (1867). As a consequent opponent of speculative natural law, he supported the view that human history is subordinated to the laws of evolution and claimed that it could be divided into different developmental stages. This idea of developmental stages and developmental principles of law determined the comparative legal science, as it was presented by the founders J. Kohler, F. Bernhöft and G. Cohn, and especially on the pages of the *Zeitschrift für vergleichende Rechtswissenschaft* (1878). Kohler, using Post's views as his starting point, maintained that the comparative ethnological method has not led by chance to the discovery of striking parallels between the legal practices of all peoples. Rather, these practices have to be seen only as emanation of the common human nature (Post, 1895, p. 4). The basic idea here is that the social and legal institutions are defined by the same biological, psychic and social dispositions in humans and as such have necessarily to cause the same results.

However, the critics of contemporary comparative law not unjustly claim that this type of legal ethnology ignores the cultural-specific differences, establishing instead mechanistic and deterministic regularities.

This way, the historical and cultural distinctions of legal reality are replaced by the abstract and imaginary uniformity of a general theory. One might suspect that the idea of human freedom is thus denied together with the concept that human history is not a strictly foreseeable process but a development with open possibilities (Constantinesco, 1972, p. 157).

This critique may be justified (if at all) in viewing older versions of evolutionary theory as 'developmental mechanics'. However, the modern theories of pre-biotic and organic evolution are far away from such a determinism of the laws of development. Therefore, the notion of a determined by regularities and unavoidable (despite all relapses) progress of humankind has changed as well. In this sense, the application of the biological theory of evolution to the socio-cultural sphere has acquired quite an ambivalent character.

2.2.4. *The Ambivalence of Evolutionism*

19th century evolutionism, which appeared before Darwin, was a contribution to the Enlightenment and rationalism for it included a theory of a based on regularities progress in the history of humankind in general and the history of law in particular. Darwin himself, who explicitly rejected the "Lamarckian nonsense of progress" in the evolution of organisms, however believed that in the sphere of 'socio-cultural' evolution there is advancement and progress. With this idea he even offered consolation to all those who were taken aback by the view of the "descent of man from a lower organized form of life". In fact, according to Darwin, the fact that man has gradually reached the highest level of organic development and has not been placed there from the beginning on, gives us some reason to hope that the future will bring him even greater achievements. (Darwin, 1975, p. 380)

Darwin's contemporaries and advocates shared his opinion with regard to the progressive development of humans and tended to embrace an absolutely optimistic view. Yet, Darwin himself tried cautiously to avoid any naive ideology of progress. The application of the theory of evolution to human history doesn't necessarily include the assumption that knowledge, morality and religion develop in a progressive manner.

Rather, Darwin was confronted with enough examples, showing that "knowledge gaining in humans" can be "the reason for a temporary but long lasting moral degeneration". An anonymous critic claimed in *Spectator* of March 12th, 1871, this position to be a "new theory of the fall of man", which was not clearly recognized even by Darwin himself – in this respect evolutionism seems to justify the Jewish-Christian doctrine of the moral decline of humans, caused by their search for knowledge.

However, modern paleoanthropology has made another decisive step by rejecting the myth of man's original innocence and by claiming that already our ancestors – *Australopithecus, Homo erectus*, and Neanderthals – were cruel enough to commit murder and to practice cannibalism. Some radical anthropologists even hold the view that *Homo sapiens* has proved to be the most prominent killer in the animal kingdom in the course of a murderous inter-generic struggle.

Indeed, there is convincing evidence that the earlier presentations of human evolution as a progressive line of hominid species are wrong. Rather, some of those species existed simultaneously (See Oeser, 1987, pp. 44-46, 1988, pp. 158-160).

From this point of view the old theories of man's originally carnivorous nature have been revived. It is enough to remember Hobbes' *homo homini lupus* – a metaphor, which has found its confirmation also in the most recent historical events – the two homicidal world wars with many millions of killed people.

Thus it doesn't come as a surprise that in the contemporary literature on evolutionism the gloomy theories about degeneration and decline of humanity prevail. This shows the ambivalence of evolutionary theory, which with its two basic factors – mutation and natural selection – represents a mixture of chance and necessity. The theory can therefore be used as an explanation of reason in human history as well as the opposite: A theory of human basic irrationality. Contemporary evolutionism contains both versions.

A remarkable, though remaining practically unnoticed attempt (See Zemen, 1983, p. 46) to apply evolution to law has been made by the Austrian dogmatist of civil law Franz Gschnitzer. He presented his view in a paper delivered at the *Wiener Juristischen Gesellschaft* in 1946:

> When we speak in jurisprudence of phylogenetic ancestors, archetypes, mutation, natural selection, adaptation, senility, then only in a metaphorical sense, for all these processes occur solely through the intervention of man who inherits or creates the institutions of law. We are confronted not with inheritance, but with tradition. Yet the total result resembles a biological process of heredity. I believe that I have found the explanation: Man is shaped according to natural laws and principles, and it is not astonishing that everything he creates follows the same laws and principles (Gschnitzer, 1946, p. 508).

Like so many attempts at a 'naturalistic' interpretation of the origin and development of law, this one too gives credit to the pioneer work of Ihering (Wieacker, 1967).

Gschnitzer gave rise also to the idea that it is not the legislator who determines the development, but vice versa. This view was emphasized by Hayek who claimed (in a somewhat extreme manner) that cultural development was essentially an unconscious process: The individuals were selected by chance or fate – like in the organic evolution. Hayek offered not only the already mentioned historical argumentation of the evolution of morality and law, but also a systematic one – by his theory of complex systems. However, he didn't regard Darwin or other 19th century evolutionists as the true founders of this theory. Rather Hayek referred to Adam Smith whose view of a self-ordering process, creating structures like an "invisible hand" he considered as the basis for a theory of cultural evolution.

In order to overcome this mere metaphor (which in Smith's work has additionally a quite mechanistic character), Hayek (1983, pp. 169-171) established his own view by introducing a notion of systems, thus complementing the classical two categories of artificial systems which are entirely the result of human intention and natural systems, which are completely independent of human planning and action with a third one: Systems which are indeed the result of human action but not the result of human intention.

The third type of systems includes complex social systems like morality, law, economy, language and science and is subject-matter of evolutionary order. An external feature of such systems is the fact that none of their members can possibly understand the systems so profoundly as to be able intentionally to alter them.

But what does complexity mean? While Hayek himself doesn't give a precise definition, one can nevertheless maintain that complex systems are indeed organized in such a way that no external or internal observer could reach to the bottom of them. Still, as the following diagram shows, the structure of complex systems and, thus, the notion of complexity is amenable to a rational explanation (See Oeser, p. 169).

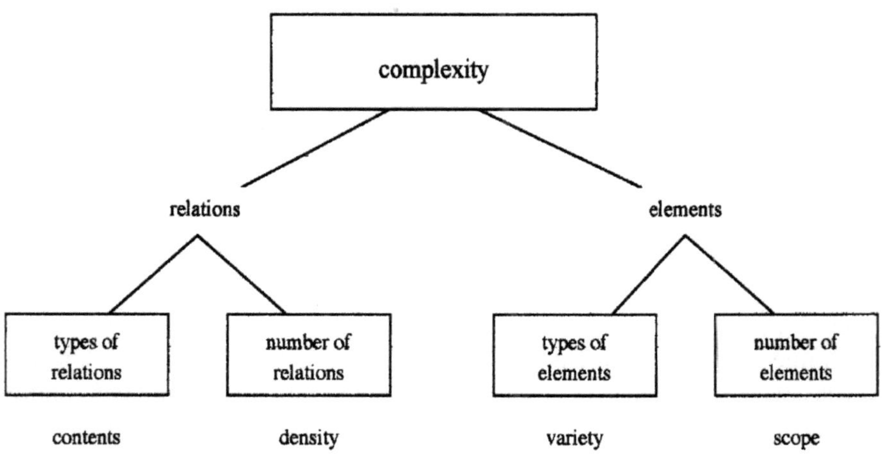

This diagram should basically demonstrate that the concept of complexity is not – as frequently claimed – the magnitude of a system but a notion of the diversity of elements and their possible and real interrelations, which reflects precisely the character of human institutions like morality, law, and economy: They are not simply quantitative phenomena but are rather to be characterized by specific qualitative properties. This would mean that one may assume evolutionary stages in human history, which manifests itself in the increasing complexity of its institutions without assuming deterministic developmental laws, constraining this history. It is important to keep in mind that the increasing complexity has to be always understood as a qualitative difference, which can completely change the structure and the properties of a system.

To assume evolutionary stages in the development of human institutions would mean to assume an order existing before the constructive and planning reason. In this aspect one has to agree with Hayek who remarks:

Not intelligence is the source of order, but order is the source of intelligence. The latter is a sub-order that reflects the characteristics of the more comprehensive order ... order doesn't presuppose reason as claimed in animistic and anthropomorphic views of the past, but reason itself is one of the developed, ordered structures, providing any human being with a kind of a model or a map of the world in which he or she lives, and helping him or her to interpret different events and to adapt his or her reactions to one complex structure of alternative expectations of things to happen (which vice versa follow his or her different reactions). Once formed, the self-replicating order increases the possibilities for the development of more orders of this type by the way of sustaining and promoting the multiplication of constructing sub-orders, thus, bringing to particular objects that contribute to the further development of the universe, where they are placed (Hayek, 1983, pp. 181-182).

At the same time one should not deny the possibility that the constructive and planning reason – at a higher level of insight – can intervene, at least locally, with the complex system. Only through this intervention of the planning reason evolution of nature turns into a history mankind.

2.2.5. *Evolution and History in the Contemporary Discussion*

Before attempting to shape a developmental theory of law it is necessary to clarify in the contemporary approaches to the relation between evolution and history, which prove to be quite controversial.

Ever since Dilthey a clear distinction has been made between natural history and the history of mankind:

We explain nature, but mental life we understand.

This view has led to a methodological barrier between the study of nature and the study of the human history which – the already mentioned conceptions are an exception – seems to be still valid today. This dogma in philosophy of science actually prohibits any unifying evolutionary view of nature and culture, the latter appearing not related to each other at all.

The alternative extreme however is even less acceptable. It would mean that human or cultural history is to be identified with natural evolution or to be looked upon to as a part of the latter. Such a view can have – as social Darwinism has shown – devastating consequences.

Hence, the question arises whether cultural history or socio-political history of humankind would not be enough for the establishment of a developmental theory of law. The answer is 'no!'. As Habermas has already shown, mere historical descriptions have basically a "narrative form". They are constrained by the referential system of narrations and as such cannot lead to any developmental theory. In the best case they provide for the respective empirical material. Therefore, the comparative universal history, trying to establish similar processes and structures only on the basis of this descriptive material has to be expanded to an explanatory concept from the point of view of evolutionary theory.

Josef Kohler anticipated such a development. However, his starting point was different. Thus he didn't aim at an evolutionary theory of law, using the material supplied by the universal history but *vice versa* – his goal was the creation of a universal history to which evolutionary theory was applied.

> The universal history of law has to demonstrate the development of law in history. In the context of cultural history it has to show as well those cultural results that were linked to law and the influence of law on cultural life and cultural progress. At this stage, legal philosophy will have to demonstrate the role of legal structures in the teleological development of world processes and ist role in the future of various cultural systems. Also, legal philosophy will have to make clear the goals of humankind under the protection of legal order and the contribution of law to the final destiny of the world. Without universal history there is no adequate legal philosophy, and without universal history there is no philosophy of humankind (After Zemen, 1983, p. 2).

What this view calls to life is a hybrid form between a narrative presentation of history and retrospective explanatory evolutionary theory, characteristic for the speculative philosophy of history (e.g. Hegel). This methodologically untenable 'mixture' was opposed by Habermas who maintained that the *aporiae* of philosophy of history had appeared from the attempt to transform the developmental theory into a universal historiography. Evolutionary theory should not be supposed to play the role of a theory of history, for history has no capacity for theory. Evolutionary theory cannot be applied to historiography, but rather to practical discourse. (Habermas, 1976, p. 313)

However, as Habermas himself stated, this cannot mean that the (socio-cultural) theory of evolution – because of its importance to historical research – should not be attested any meaning for history as historiography. According to Habermas, evolutionary theory has more or less two functions in historiography (and thus in legal history):

- Evolutionary theory could be used as a meta-theory in order to evaluate competing narrations in the same sphere of phenomena and to stimulate the "discourse development of tradition" by historians. This actually has been always an object of interest of the so-called 'historic'.

- The reflexively applied evolutionary theory can explain "developmental thrusts, making epochs", and explanations can become a part of a rational discourse of a time-diagnostic theory of the present: Time-diagnostician of this kind (e.g., Freyer, Marcuse, Bell, Dahrendorff) take the fictitious perspective of an evolutionary explanation of a past lying in the future.

Thereby, with regard to the explanatory claims of evolutionary theory, Habermas goes beyond Niclas Luhmann, who reduced those claims to a mere "study of possibilities".

Luhmann, explicitly attempting an evolution of law (1970, pp. 3-22) as early as 1970, had to interpret the explanatory possibilities of the theory of evolution in a very restrictive manner, his own starting point being a systems theory within the functionalism of social sciences. In fact, he radicalised the classical approach of Talcott Parsons. Parsons had done the step towards a general systems theory of society: He had replaced the anthropological and historical data and history by systems development, already perceiving development as an increasing complexity of systems. However, he claimed a difference between the sphere of objects and systems-theoretical statements and conceded that the cultural interpretation system has a causally independent status within the sphere of objects in society. On his part Luhmann emphasized the meaning of a systems theory of the social sciences so strong that he applied it to himself in a self-referential manner. This radical view of systems theory is supported by the biological theory of autopoiesis as conceived by the neurobiologist Maturana. Luhmann (Riedl, Kreuzer, 1983) resorted to this theory in 1982 at a Darwin Symposium, and this – strangely enough!- without mentioning its founder. Since, however, evolutionary theory is used only within the conceptual frames of the systems theory, the shortcomings of the systems view in social sciences are not eliminated, but rather replicated.

Habermas (1976, pp. 335-336) – in his reply to Luhmann's offer (1976, pp. 284-286) for cooperation between sociologists and historians – clearly referred to these shortcomings of Luhmann's systems-theoretical functionalism: It can neither explain the emergence of new structures, nor can it determine the identity of social systems in the course of their changing structures and states.

Luhmann himself recognized the shortcomings, which constrained his view. He stated that evolutionary theory in his perspective is neither a "process theory" nor a "developmental theory". According to him, the theory of socio-cultural evolution doesn't deal with processes, i.e. with changes at any given time – from t(1) to t(2). Also in the explanation process it cannot "move" to the beginning: It has to do without a "particular beginning".

Hence, the theory had to abandon important explanatory claims, which are inherent to evolutionary theory in contrast to a merely descriptive natural history. Following Luhmann, one should be prepared to accept such limitations in a theory of socio-cultural evolution. Accordingly, its analogies to the explanatory structures of organic evolution are rather abstract:

- Language represents the mechanism of variation,
- Communication media represent the mechanism of natural selection,
- Roles or systems differentiation represents the mechanism of (re)-stabilization.

It is clear that this way the transfer of the explanatory structure of the biological theory of evolution to socio-cultural evolution has reached not even the level of analogies, but rather the one of mere metaphors. In particular, this can be seen in the very hint at language as mechanism of variation, which, as Habermas critically emphasized, can hardly adequately represent the "bio-chemically analysed process of mutation". (Habermas, 1976, p. 337) What is required here is a genetic theory of language-mediated cognition in the domain of objective scientific thinking as well as in the sphere of moral practical insights. Such a fundamental evolutionary theory of human cognition and morally relevant action is given in different well-developed conceptions of evolutionary epistemology (See e.g. Lorenz, Vollmer, Riedl, Wuketits, Oeser) and ethics (Mohr, 1987). The pioneering works by Lorenz to which almost all of these conceptions refer, have already left an impact in jurisprudence. Apart from the evolutionary views in social science, Herbert Zemen in his *Evolution des Rechts* (Evolution of Law) explicitly refers to Lorenz' *Die Rückseite des Spiegels* (Behind the Mirror), and tries to connect it with the writings by Gschnitzer, Kohler and other legal theoreticians like del Vecchio (Zemen 1983).

Another fundamental approach, which developed in economy and expanded to legal science, is the evolutionary understood theory of complex systems by the already mentioned legal scientist and economist von Hayek. However Hayek's basic idea is that complex systems initially created by man, because of their complexity can be no longer fully understood by any human being. Each intentional rationalisation remains always incomplete. But such a rationalization is – and this is the decisive point – unnecessary. Like all human institutions developed in the process of evolution, morality and law do function even if none of the members of a given society fully understands the behavioural rules according to which he/she acts, and even if nobody knows how these rules have developed. It is the system itself that functions, or, to use Hegel's phrase, it is the "cunning of collective reason" which governs the behaviour of the individual, without the latter being aware of it (See Oeser, 1988). Yet, also Hayek cannot deny the existence of a general evolutionary law, and he indeed repeatedly and almost literary, refers to Spencer's great idea: The formation of ordered structures presupposes the divinity of interrelated units as well as the stronger coherence of some of the units... Greater order (and heterogeneity) will be created by better chances for endurance (survival) in those species that are endowed with a still better capacity to maintain or to restore their coherence (Hayek, 1983, p. 177-178).

Needless to say that these considerations do not only refer to Spencer's old evolutionary law, but are also hinting at the modern theories of self-organization (Haken, Eigen, Prigogine). Moreover, one can assume a fundamental congruence between all these views and the idea of universal evolution as an iterative process (See Oeser, 1987). Hayek too stated that such an integration of elements into complex structures as it happened in the process of the origination of life (primitive elements together built more complex systems), had to be assumed "also at the level of more complex structures". He maintained that consequently, a cumulative process would bring forth a hierarchy of structures. Those structures that exhibit greater coherence or flexibility show also the tendency to protect themselves effectively against destruction. The advantage of an element, becoming a part of a complex system will therefore be its increased probability and progressive replication. From the atom, protected by the molecule, the cell, protected by the organism to the individual, finding security in the society, there is a hierarchy of superimposed orders. Their preservation by the means of a process, which among all random variations selects those that have found the protective shell of a bigger structure, can be sufficiently explained (Hayek, 1983, pp. 178-179).

However, it doesn't matter if the question how such structures are created remains unanswered. "Protective shells" of single elements anyway cannot exist before the existence of their contents. Thus, one has to postulate for each structure its own mechanism, which is just given by the iteration of flowing and structural information.

This is to say that elements build their "protective shells" or their system's borders by the quality and quantity of their interrelations. They stabilize the initially flexible relations and form structural patterns that can again be parts of greater systems of order. With regard to socio-cultural evolution, the structural conditions of the human brain as a product of organic evolution lead to a very similar situation. In the course of millions of years neuronal-anatomical structures have developed. They determine our thinking, decision-making and acting as iterative processes leading to more and more complex constructions. It is therefore not necessary to look for a third 'mediator' between the pure constructive reason and the mere intuition in the sense of Hayek's interpretation of Adam Smith's "invisible hand". It is rather the practical reason as primacy in Kant's understanding that has to cope not only with the changing conditions of the natural environment but mainly with the changes in the social environment, on which humans depend more than all other living beings.

2.3. *Approaches to an Evolutionary Theory of Development of Law*

Within a comprehensive theory of socio-cultural evolution, any developmental theory of law is an attempt to represent – by means of rational reconstruction (which contains descriptive elements from the comparative legal history and requires normative elements from ethics and logic) – the progress of the idea of law, aware of the existence of human freedom. Essentially, the idea of law is no longer regarded as a static phenomenon in the sense of a 'eternal truth' but as a developing phenomenon.

The just law itself is a dynamic concept, which is compatible with the nowadays not to be any more denied idea of historicity of law.

Unlike any historical relativism, a developmental legal history necessarily maintains at the same time the existence of a developmental rule. There is no theoretical explanation apart from a subsuming an event or a case under a lawful order. Thus, a developmental legal theory basically consists of the arrangement of one or more fundamental laws of development that determine the events. Yet this doesn't mean – as it is almost always claimed (e.g., Hayek, Habermas) – that these laws are deterministic.

History of science – especially after the eclipse of deterministic physics (Laplace) – should have made clear that in no sphere of human history there can be something like deterministic laws, helping to precisely predict the future. The myth of Laplace's ghost had to decline at the latest with the foundation of quantum mechanics, the preparation for which had started much earlier with thermodynamics (See Oeser, 1988, p. 82).

Empirical natural sciences and empirical human, respectively, social sciences have left space only to statistical laws or laws of developmental stages in the sense of increasing complexity with non-deterministic and therefore non-predictable transition phases.

Such a view is required by an evolutionary theory of law which, (making use of material from historiography of law) methodologically reconstructs phases of the development of law, in order to represent (by the example of law), the progress of mankind under the auspices of freedom, i.e. self-determination.

A developmental theory of law has to adequately answer two requirements:

- It has to be able to reconstruct the transition from man as a natural being (analysed and explicated in terms of biology) to a man as a legal being (explicable no longer in the terms of biology).
- Moreover, it has to be able to reconstruct in a non-biological and non-empirical (different from natural sciences) way the evolutionary stages of law.

Now, what is the type of the non-natural-scientific level of explanation? It is not the level of evolutionistic and systems-theoretical legal sociology (see again Luhmann). As we have seen, this type of legal sociology remains at the level of metaphors and goes hardly beyond analogies. The same is true to the "second", non-Darwinian conception of socio-cultural evolution, which is just imitating organic evolution of Hayek. To attain to the third stage of the application of the evolutionary concept to the phenomenon of law, i.e. the level of isomorphy or homomorphy, one requires a logic of development or of processes. By means of such a logic, the developmental processes of law in the sense of their claims for truth (values or other ethical principles) can be reconstructed. This logic of development is a deductive, statistical conception, which transgresses the merely formal logic of statements (sentences) (See, in particular, Habermas, 1976, pp. 338-340). It means, according to Habermas, "collective structures of consciousness" in the sense of structural possibilities of levels of social learning. Such a logic of development was anticipated in the notion of dialectics in the philosophy of history of German idealism in the spirit of Kant. Habermas however is warning – correctly at that – of a hasty "Hegelian connection" between the logic of development and history, within a theory of the objective mind.

One cannot but agree to Habermas' idea that there exist two interdependent (mutual dependence), but opposing causalities:

- Causality as structural possibilities or universal conditions of possible learning processes in the sense of structures of consciousness.
- Causality as factual causality between events, occurring as causal processes.

The connection between these two types of causality has been discussed in biology as the "principle of emergence" which is supposed to explain the appearance of new system qualities on the basis of the 'cooperation' between the downward causality and upward one (See Oeser, 1987, p. 77). Upward causality is to be understood as the usual connection between real events; downward causality means the "structural" causation, which can be represented by organismic structures (e.g. brain or genome), as well as functional structures or patterns.

Habermas stated that the connection between these two causalities – that help us to explain the effects of structure upon the history of events and, vice versa, the effects of historical events upon structures – could be seriously analysed "only within a logic of processes and the solution of problems" (Habermas, J., 1976, p. 341). In addition, one should say that such a logic of processes is documented by the biological evolution. According to the common view of today, experimentally supported by science, any organism is – functionally seen – nothing but a problem-solving machinery that increases its own functional capacity and complexity by means of self-organization.

However, to apply this biological view of self-organizing increasing complexity to an evolutionary theory of development of law without ending (as the already discussed models of evolution of law) on the level of metaphors or analogies, would presuppose a fundamental differentiation.

2.3.1. *The Two-Stages Concept of Evolution of Law*

Any application of the concept of evolution to a developmental theory of law requires the conception of levels including necessary and sufficient conditions:

- The natural evolution of living beings including the evolution of hominids is to be regarded as a necessary condition and a first step of human history. Without this process the biological species *Homo sapiens* would not exist.
- As a sufficient condition of human history with its cultural development in the spheres of economy, morality, law, arts, and science, one needs additional mechanisms, namely, the activities of human consciousness or reason. Without them one cannot imagine any development beyond mere biological survival.

Such a two-stage concept avoids any reduction of cultural history to natural evolution as well as the belief that there is no relation between both spheres.

A concrete and widely accepted example to be already found in the juridical literature (see Naucke, 1986, p. 53-54, Bydlinski, 1988, p. 145) is the incest taboo (the prohibition of sexual intercourse between kin group members): It is a part of the normative order of numerous societies, past and present. Juristically or from the point of view of legal philosophy there are only weak reasons for incest avoidance. The question is, why in any concrete case incest should be regarded illegal. The answer comes from biology, i.e. from human genetics. There is evidence that the offspring of incestuous reproduction tends to genetic diseases. For some jurists this scientific founded fact suffices as a basis for a legal prohibition of incest. According to Bydlinski we do not necessarily need other deeper reasons for this prohibition. (Bydlinski, 1988) Some, however, tend to be more cautious. Thus Naucke asked: Why shouldn't be one's offspring handicapped? There are many answers to this question none of them seem to be more valid than the others (Naucke, 1986, p. 54).

An answer may come from our empathy with the person exhibiting genetic disease. Another answer could follow economic reasoning (handicaps are a burden to healthy people), or one might generally argue against a genetic deterioration of humankind.

None of these answers taken separately for itself supplies the absolute solution of the legal foundation of incest avoidance. In particular, the quite plausible argument against (genetic) degeneration of humans is too general. Actually, each medical treatment acts against natural selection. Such a treatment can preserve the life of people who otherwise would not survive. What is needed is a multi-level structure of explanation. The scientific argument from the fields of genetics and evolutionary theory is not sufficient even in such seemingly plausible cases. The foundation of moral or legal norms through the obligation to take care of handicaps or sick people requires a level different from the biological theory of evolution by natural selection. Any attempt to eliminate these differences leads to an unwarranted reductionistic biologism.

The failure of biologism – which being biologically unjustifiable is usually not advocated by biologists – is the imagination that human history is nothing else but the linear continuation of natural history (= genetic, organic evolution). In fact, since the formulation of modern synthesis (see Huxley, 1942) – the synthesis of genetics and descriptive phylogenetics – the explanatory system of evolutionary theory is a multi-level model that includes arguments about the changes or the evolution of the evolutionary mechanisms. If one accepts the notion of socio-cultural evolution, then one is compelled to see that at this level different evolutionary mechanisms are at work and that there is a significant acceleration of the evolutionary changes. Organic evolution with its own and different mechanisms occurs at much slower pace (See Oeser. 1987, pp. 21-23, p. 189).

In the written history of humans – during the past 5000 years or so – there have been almost no genetic or organic changes. The comparatively low pace of organic evolution makes clear that biologically explainable processes are governed by their own mechanisms.

2.3.2. *The Transition from Natural State to Legal State*

We have preserved by constant reproduction the same brain, which long ago worked in the skulls of barbarians and savages. All important features of humanity have developed out of the scarce seeds of thinking in its earlier days.

This statement had been made by Morgan (1871) in the year of the publication of Darwin's *The Descent of Man.* It reflects Morgan's attempt to establish an evolutionary view of man's early societies. Today we are quite well informed about hominid evolution, which has been, in the first instance, a development of the brain's structure and size. We can say that during the comparatively short time of pre-history and written history of mankind the biological equipment of the brain has remained unchanged. Thus, the 'natural state' to which so many political and legal theories refer, is, from a biological point of view, no longer to be regarded as an enigma. We can correctly assume that the genetic make-up of the *Homo sapiens* has not changed since the Neolithic revolution, when the early societies were formed. However, this had been already known by the old theorists of evolution at Darwin's time. The classical formula in this respect comes from Wallace: His mental capacities enable man to keep his unchanged body in harmony with the changeable universe. (Wallace, 1864) This means nothing else but man's 'universal' capacity for learning. This unique for the humans is a property of the central nervous system that is a product of organic evolution. Hence, our cognitive apparatus is constrained by genetic factors, which, however, are not non-surmountable and therefore are to be seen as "genetic propensities" or "inclination structures" (Mohr, 1987, p. 50). Genetically constrained patterns of behaviour that are governed by the central nervous system are not just life-preserving functions – such as locomotion or the seizing or catching of objects – and not only all those patterns of communication (smile, brow, etc.) that have been studied by human ethology. They also include emotional and moral structures.

Especially Hans Mohr (1987, pp. 76-78) has shown that in the sphere of human emotions in particular there exist genetic propensities, which – like our innate cognitive structures – constrain socio-cultural evolution as well as the ontogenetic development of any individual. However, in contrast to all other animals endowed with a central nervous system, human ontogeny seems to be much more important than phylogeny (Oeser, Seitelberger,

1988). Initially, any individual human being starts its life where the genetic evolution of hominids ends. This means: In the short period of some 5000 years of the very human history – a period, which is rather insignificant in the geological and evolutionary time-table – there have occurred, as already stated, no genetic changes. In other words, our emotional and cognitive make-up is basically the same as it was in the late Pleistocene. An example is our tendency to think in terms of linear causality without feedback. This tendency finds its expression in our destructive, exploiting behaviour which characterizes the mentality of nomadic small groups that lived on wide territories and could make use of plenty resources. Until recently, this mentality has been very typical of humans. Today however, we must face the trivial fact that our natural resources are not unlimited, so that we have to learn how to change our behaviour. The same is true to our legally relevant structures of moral conduct.

These structures are adapted to the requirements of small societies which, in former times occupied and exploited the ecological niches. In such societies a morality of small groups developed. It includes indeed simple, but very strictly executed legal-like rules that are recorded in the individual's consciousness. These rules and forms of behaviour aptly regulate the survival in a small group living in rich environment, but, at the same time, also competing with other such groups. The result is what socio-biologists call *inclusive fitness.*

The fitness of any individual is therefore not to be measured only on the basis of its own genetic and mental outfit (*individual fitness*), which guarantees its own reproduction, but also depends on the promotion of the *fitness* of its genetic relatives or members adopted through marriage by the same group.

From this point of view, we gain an explanation of the human natural state and are not forced to advance further sociological or social- or legal-philosophical reflections. The natural state of man is not to be regarded as completely free of legal norms, as Hobbes' metaphor *homo homini lupus* seems to indicate. An evolutionary theory of law doesn't recognize an absolutely free of legal rules state (see Wesel, 1984), but rather only different developmental stages of law. Morality and law exist from the very beginning of human development and were a basic condition of hominisation. The first human being was, as it were, a group with internal rules (which regulated the behaviour of the individual toward any other member of the same group), as well as external principles (which regulated the individual's behaviour toward members of other, competing groups).

All characteristics of early or archaic law can be deduced from these basic rules. Archaic law rests on principles of revenge and defence. Defence

is connected with solidarity among one group's members, revenge on the group's enemies, and compensatory penance through the group (See Trimborn, 1950, Eder, 1980, p. 69).
These represent the behavioural patterns of a group competing with other groups. Inclusive fitness means peace within a group, aggression toward other groups. The increasing density of population on a given territory most probably supported aggression and violence.

Therewith the evolutionary origins of law as a self-chosen coercive structure become apparent. This structure itself is, on the other hand, a necessary consequence of the increasing size and complexity of human societies.

The very legal state, which transgresses the kinship system of small groups begins with rules concerning integration and marriage. Kinship and marriage are the two fundamental mechanisms of the juridical fixation of membership and the resulting rules.

Incest taboo, rules of exogamy, payment for the bride, etc. are the first practical principles at the lowest level of early law. Legal anthropology and legal ethnology reconstruct such principles. However, these disciplines presume at that a similarity of problems between today's primitive societies and prehistoric groups. According to Robert Lowie the similarity between contemporary savage societies and prehistoric ape-men is an influential doctrine the failure of which has to be underlined. This doctrine doesn't take into consideration the fact that even the simplest recent horde has a long history and has moved far from the hypothetical initial stage (After Wesel, 1984, p. 527).

However, there are at least some hints at the existence of legal rules in illiterate societies at later stages of civilization. Homer (Wolff, 1952, pp. 1-29, pp. 157-163) mentioned payment for brides in the form of gold, silver, cattle, and the *coemptio* – the buying of a woman at her marriage – was known to early Roman legal system. Yet, that had already served a purpose different from the purchase of married state (*Kaufehe*) in early societies (See Jörs, Kunkel, 1949, Wesel, 1984, p. 530).

Hence, it should become obvious that the classical view of Hobbes (Hoerster, 1979), Locke (1967) and Rousseau (1948) is a mere fiction. They believed that humans in their natural state existed as completely free, isolated individuals without any legal norms. Necessarily then, the model of social contract as an explanation of the transition from natural to legal state cannot be accepted without critique.

This criticism started with Hume who regarded the social contract as a useless fiction (See Hoerster, 1979, pp. 153-155). Thus, no state in the world is known to be founded on the basis of a contract with its citizens.

Alternatively, the utilitarian view perceives the state as an instrument of common use, which in the long run should be of use to any individual. According to its epistemological attitude toward the problem of causality, utilitarianism has considered the rules for establishing public welfare as traditional habits, bringing to the loyalty of people to their political ruler.

Modern views of such a gradually developing common law can be found in the evolutionary conceptions, e.g. the one of Hayek, who – in order to disconnect socio-cultural from organic evolution – was actually forced to resort to the notions of tradition and habit.

It is not our intellect that created our morality, which, regarding its effects, we still do not understand; rather, the cooperation of humans as dictated by morality made the emergence of reason possible... This had been explicitly stated already by David Hume – our morality has not been invented by human reason (Hayek, 1983, p. 183, p. 185).

The question arises whether the social contract model in any form is to be dismissed by a theory of legal development. Or to put it this way, has utilitarianism outlived its old opponent. Taking evolutionary perspective seriously, one may indeed notice the close relation between this pragmatism and utilitarianism and the Darwin-Spencer principle of the 'survival of the fittest'. But Rousseau and in particular Kant have outlined sophisticated social-contract theories that are well compatible to the evolutionary approach. Thus, Rousseau advanced the view of a society composed of equal members who meet and gather to find common solutions to their problems. However Kant claimed more rigorously and realistically the generalization ability of individual action, and - on its basis and as a kind of 'logical coercion' – he prohibited the violation of individual freedom.

Common freedom, common equality in the eyes of the state's authority, and autonomous self-realization of the individual are to be put together in a logical-aprioristic way. The social-contract model, then, means nothing else but the gradual insight into the necessary relations between freedom, individual, and society. A modern version of these speculations is Rawls' *Theory of justice*, (Rawls, 1975) considered by Dreier as a procedural interpretation of Kant's categorical imperative (Dreier, 1981, p. 300).

The practical realization of this insight in real history is to be achieved only in stages. It could be linked as well to an evolutionary theory of the developmental stages of law.

2.3.3. *Developmental Stages of Law*

Kant's starting point is the thesis that man as a rational being is to be defined by his attempt to achieve perfection. This achievement, however, is possible only on the basis of a reasonable socialization, bringing humans from the natural to the civilized state.

The supreme principle paving the way to the civilized state is from the categorical imperative – that one should act according to the maxim one wishes to see as a general law – deduced legal principle, namely that one has to act in such a way that his/her free will could, together with the freedom of everybody fit into a general law (Kant, 1968). The supreme legal principle aims at creating the civilized state which, in its turn, gives the real meaning to the necessary idea of pure reason of a unified legislative will, typical for each human being. Thus, an ethical theory of action (categorical imperative) develops into a normative theory of legislation, which, according to Kant has to supply positive legislation with "unalterable principles" (Kant, 1968). Contemporary equivalent to this theory is the theory of basic rights, i.e. of those fundamental principles which are necessarily present in every constitution, representing the essence of the state constitution and, thus, according to Kant, being "sacred and irresistible" for everybody. Even more so, because these principles do not constrain but guarantee the freedom of the individual in the community.

Hence, Kant found the motive for the creation of a theory of the developmental stages of law in the idea of freedom.

Respectively, the development of a legal consciousness – on the level of both, the individual and humankind – could be traced (in Kant's perspective and from the viewpoint of an evolutionary legal theory) through the following stages:

- First is the non-reflected stage of the free will as an abstract ability of 'wanting', which, being not necessarily rational can lead also to injustice. This stage applies to the archaic state of man who, capable of free will and thus of a sense of responsibility, recognized his uniqueness among the other living beings.
- The second stage represents the one of the free choice. Man is able to consciously reflect upon himself and upon the motivation behind his actions. In this he proves to be both genetically (on the basis of drives and instincts) and rationally determined.
- The third stage is the one of the free, determined by reason will. Here the individual recognizes the reasonableness of the existing legal institutions as structures chosen by himself.

These developmental stages are *a priori* pre-supposed in the spirit of Kant and reveal a logic of development which can be rationally reconstructed.

They are empirically correlated to real history (See Eder, 1980, pp. 158-160).

- The first stage of the free and abstract 'wanting' is observed already among the early illiterate segmental societies. Those societies knew no fixed authorities, but only the mutual aid among the members of a kin group. They solved the conflicts with competing societies by means of wars. The rules of legal action were tied internally and externally to concrete forms of action.

Any notion of norms remains directly linked to an immediately perceivable factual conduct – to 'pictures of acts' that gradually find their expression in particular types, disregarding the single cases. Sometimes they are 'caught' in words and sentences and become part of the tradition (After Eder, 1980, p. 162).

Already at this stage, humans succeeded in some situations to institutionalise a particular way to resolve conflicts – that ability had been improved with the increasing complexity of archaic (especially Neolithic) societies. However, the creation of law and legal action are not yet to be differentiated from the concrete context of interaction. At the same time, this context allows only for legal principles, which grow naturally from the practice of personal defence against various claims – a pay-back and revenge. Therefore, this first stage of legal development is identical with the so-called natural state of humans. Obviously, it is not an absolutely unlawful one.

- The second stage of legal development coincides historically with the civilizations prior to the modern age (Eder, 1980, p. 165). Here, law is delegated to a sovereign, who is the representative of general norms, the latter being the foundation of jurisdiction. In the despotic forms of legal execution, the ruler is automatically a judge as well. Still, this doesn't mean that the sovereign himself personally exercises the role of a judge – the latter can be overtaken by a professional jurist.
- At the third stage of legal development – historically seen it coincides with the modern times – the ruler is no longer the legal executor, he is rather the legislator. 'The transition from 'ruler-judge' to 'ruler-legislator' is the main feature of the evolution of law in the 16th century' (Eder, 1980, p. 165). Finally, the next step represents the transition to the contemporary form of legal system, where law grows on the basis of a formally regulated legislation.

The integration of the principles of legal philosophy, respectively ethics of law, on the one hand, and the empirical systematisations of history of law, respectively of comparative law and legal anthropology – on the other, would be incomplete without reflection on the methodological aspect. Thus the evolutionary theory of the developmental stages of law remains a 'blind' speculation – although supported by historiography of law and comparative jurisprudence – without the specification of the methodological mechanism for their distinction. Like the theory of evolution in general, any evolutionary theory of law in particular includes a mechanism, the principles of which serve as a basis for the explanation of real history. Here, the mechanism comprises of the juristic methodology of practical argumentation, bringing to the decision-making in each concrete case. Evolutionary theory finds its application not in historiography but rather in practical discourse (Habermas, 1976, p. 313). If, finally, one tries to summarize the development of morality and law together, one would attain at the following rationally grounded assertion: Morality and law have been mutually related at each stage. This means that the moral sense improves on the basis of legal progress, and *vice versa* -this progress depends on the 'moralization' of law, which moralization by itself brings law to a higher developmental stage, increasing its problem-solving abilities, demonstrated in more complex cases.

3. THE THEORETICAL MODEL OF A DEEPER ARGUMENTATION

For a legal philosophy, understood as a theory of practical reason, the juridical theory of argumentation seems to be fundamental and indispensable. This, however, couldn't get the consent of all trends in legal philosophy. Thus, all forms of objectivism - be it the based on ethical values, ontological one, or, the naturalistic one – as well as all forms of subjectivism, intuitionism, and emotivism are practically irreconcilable with the fundamental justification claim of argumentation theory.

It doesn't really matter whether a legal decision rests upon immediately comprehensible objective values and objective principles of legal ethics, or upon subjective intentions and emotions. In all cases, one presumes the *a pripori* existence of evidence, to be somehow always recognized and experienced, which then serves as eternally valid ground for decision-making. However almost all such efforts – both in the past and, even more so, in the present (see the concept "nature of the things") – have proved to be either devoid of substance or extremely vague. Even if it was not so, each type of justification would only be unnecessary or of a secondary importance. The starting point of a theory of argumentation, however, shows to be completely different.

Characteristic for this theory is the procedural, step-by-step justification, which remains all along unlimited. The limits one is confronted with are not theoretical but only practical in nature and must be 'pushed' whenever the real conditions of life alter. Each such shift of the limits has to be clearly argued, and not only on the basis of a deductive inference but of the whole process of the rationally justified, practical juridical decision in a particular case, including all special operations of discovery of law (juristic induction and construction).

The basic function of the juridical theory of argumentation within any legal philosophy is therefore given in the procedural justification and concretisation of abstract and formal principles of legal ethics. Legal philosophy, then, reveals itself basically as a meta-theory of juristic argumentation, which doesn't only include the internal juridical justification within positive law and its reflection, but also refers to the external justification in the positive legal system. This justification is based on principles of legal ethics, external to positive law.

3.1. *The Meta-Theory of Juristic Argumentation*

The best for now elaborated concept of a meta-theory of juridical argumentation, most adequately answering the claims of a legal philosophy as a theory of practical reason is a conception, appearing under the same title (Krawietz, Alexy, 1983) and comprising three theories:

- The theory of transformation in law by Aleksander Peczník. (1979, pp. 163-165)
- The theory of juridical discourse, oriented at the concept of practical rationality by Robert Alexy. (1978-83)
- The theory based on Wittgenstein's notion 'form of life', by Aulis Aarnio. (1979)

3.1.1. *Theory of Transformation*

The theory of transformation shows clearly that 'argumentation' should not be understood only as a deductive inference. A. Peczník perceived the concept of transformation as a 'leap' from one level of human knowledge to another. He argued this – in itself a general thesis in philosophy of science – on the basis of the classical differentiation between context of discovery and context of justification. Philosophy of science has always used this differentiation in an attempt to explain the discovery of regularities in the form of new hypotheses through 'leaps', which subsequently have to be brought in 'order' again. (See Oeser, 1976) Peczník stated himself:

> The absence of a logical conclusion and the act of 'leaping' are two sides of the same thing. The performance of 'leaps' is a psychological fact to be dealt with in the so-called context of discovery. In the context of justification, one can argue, whether knowledge obtained as a result of 'leaps' is to be justified (See Krawietz, Alexy, 1983, p. 14).

From the stand-point of the contemporary philosophy of science – as long as it has already adopted heuristic – this whole 'deportation' of the context of discovery to the sphere of psychology is no more admissible. Still, one has to agree with the fundamental thesis of transformation theory that 'the transformations could be criticized or defended, and as right or false apprehended'. This would mean that there exist not only regular acts of 'leaping' but also rules for the 'leaping', the latter being the very transformation rules. Therewith the leaps will not be – as one could assume – eliminated. The only exception is the single case where it is possible with a later date to re-form the leaps into deductive inferences. These inference rules for elimination of the leaps are analytical and concern the leaps that prove to be no real ones, but merely exist on the same level of argumentation. Real leaps could not be eliminated.

- An example for such a real leap, leading to another category level is without doubt the classical controversy between 'is' and 'ought' in legal philosophy.
- A still further category of leaps is the one, leading from the non-legal, moral 'ought' to the legal 'ought'.

However within positive law there are leaps or transformations – and quite a lot at that – which do not represent deductive inferences and which could be divided into three groups:

- The transformation of legal sources, representing a transformation of a primary legal source to a secondary one, whereby the propositions about the legal validity of the secondary sources do not follow logically from the primary ones. The list, enumerating secondary sources is open and contains among others prejudication, textbooks of jurisprudence, foreign cases, etc.
- The transformation of legal norms, which refer to normgaining from the legal sources. These norms are not restricted to mere deductions.
- The transformation of decisions, based on the fact that the decision in a lot of, if not in most of the cases cannot be logically derived together with the case description from the formulations of laws and/or legal sources. In those cases, the judge must elect a decision, which he considers to be the best one from the point of view of law. He, so to say, 'leaps' to a particular sentence, which is not simply deducible. Accordingly, the concept of the transformation of decisions remains quite vague.

Still, at the end, the theory of transformation raises the fundamental epistemological question: How is it possible to gain a justified knowledge on the basis of all these non-deductive transformations? Or namely, on the basis of transformations of facts and non-legal values to come to the legal 'ought'; of vague legal criteria – to the precise identification of law; of primary legal sources – to secondary ones, the validity of which cannot be derived from the validity of the primary sources; of sources – to norms, the latter being deduced not only from sources; from these norms – to decisions, which do not follow logically from the norms.

Alexy's theory of rational juristic discourse offers an answer to this question.

3.1.2. *Theory of Rational Juristic Discourse*

Alexy elaborates the theory of rational juristic discourse as a special case of the general practical discourse. There exist various drafts of the general discourse theory, which could be – each to a different degree – extrapolated to the specific juristic discourse, namely (See Alexy, 1983, pp. 53-219):

- The practical discourse in analytical ethics (Hare, Toulmin, Baier), and its respective linguistic-philosophical fundament (Wittgenstein, Austin).
- The consensus theory of truth of Habermas with regard to *Sprechakttheorie*.
- The theory of practical advice of the Erlangen Constructivist School.
- The new rhetoric by Chaim Perelman.

Introducing some constraints, Alexy works out all these concepts into a theory of general practical discourse, subsequently deriving from it the theory of juristic discourse as a special case. The practice to proceed from general to special is hardly common among jurists. Yet, here such an approach guarantees a particular advantage by providing an immediate access to legal philosophy, which *ex definitione* has to make the step from general principles of legal ethics to the justification of the 'morality' of positive law. Hence, the theory of general practical discourse shares the same level of argumentation as legal philosophy. The only difference lies in the fact that the theory of juristic discourse is restricted to the methodological and linguistic-analytical, respectively – to the linguistic-pragmatic dimension. To turn back to a legal philosophy in the classical sense of the word would mean to remove these restrictions.

Alexy himself characterizes the juristic discourse theory by the following theses:

- The theory of rational juristic discourse deals with one of the special institutional stipulations (binding to law, prejudication and dogmatic), namely with the special case of general practical discourse (thesis of the special case).
- It is a normative-analytical theory, which employs the concept of practical reason or of practical rationality. Its goal is to ensure that the claim raised by juristic sentences can be rationally or reasonably justified, and to explicate and provide criteria for the rationality of juristic justification.
- All this takes place in the frames of a particular type of theory – a procedural one (See Alexy, 1983).

All these theses could be applied to a legal philosophy as a theory of practical reason, granted that the restriction of the theory of argumentation to a theory of linguistic-analytical discourse is removed, and the normative-analytical theory is *a priori* expanded to a normative-synthetic one.

However decisive here is the character of the procedural theory, which reflects the fact that each normative proposition is true if, and only if, it represents the result of a procedure.

In a fundamental expansion of his, initially, theory of juristic discourse, Alexy differentiates between three procedures, essential for the whole legal system with its adjucative and executive organs:

- The procedure of creating law by the state [Pr],
- The procedure of juridical argumentation in a narrower sense [Pj],
- The procedure of judicial proceedings [Pg].

Thereby, Alexy linked the procedural theory to the already discussed Pecznik's theory of transformation.

Alexy himself elaborated the second procedure – the one of juridical argumentation, the analysis of which proves, consequently, to be more extensive than the analysis of the other two procedures. With regard to the first one – the procedure of creating law, he emphasizes solely the necessity of a regulated procedure of 'positivation' of law and offers different types of state created law, reaching from a "gradual development of the legal system on the basis of adjudication to the establishment of comprehensive codification by the Parliament" (Pecznik, 1983, p. 53). All these types find their complementation in the later elaborated theory of constitutional justification (Alexy, 1985, pp. 520-521), arguing the necessity of authoritative constitutional adjucation, which has to be, however, controlled. The same is valid for the procedure of judicial proceedings, which stipulates not only justification, but decision-making as well. Still, this stipulation shouldn't be interpreted as a 'farewell to reason', because the decision has to be validated on the basis of the juridical argumentation.

Even without going into details, regarding the particular rules of external and internal justification, of transformations into law and within law, it can be said that Alexy's theory of juridical discourse suits more the methodological and argumentation doctrine of a scientist in law (a theorist) than the one of a practising jurist or judge. Yet, legal philosophy as a theory of practical reason emanates rather from juridical practice – *jurisprudentia* than from *jurisscientia* (see Kriele, 1979, p. 17), i.e. it is based primarily on the model of the judge.

3.1.3. *Theory of the Common Forms of Life*

Alexy's theory of juridical discourse grows on the example of existing in the reality individuals. According to him, this makes the difference between his theory and J. Rawl's theory of justice, using constructed individuals in a constructed time in the frames of a renewed contract theory. Alexy's individuals participate in the juridical discourse with particular convictions, rooted in their respective "forms of life". (Aarnio, Alexy, Peczník) Thus, the theory of discourse proves to be dependent on a theory of the forms of life.

Such a theory had been outlined by Aarnio in the frames of the meta-theory of juridical argumentation. It is based on a theory of values, which, considering the already mentioned stipulations could be nothing else but 'non-cognitivistic' and 'relativistic' in the moderate sense of the word. Thus if there were absolute, directly detectable values, each argumentation would have been superfluous.

According to Aarnio's theory, values couldn't be "fully justified by a reference to objective facts". (Aarnio, Alexy, Peczník) They are rather attitudes or a matter of personal taste:

> To a certain extent it is possible to use them as arguments. It is possible to ask continuously why something is this way and not that way. At a certain point no further answers could be provided. One reaches a limit, beyond which there exist no more possibilities for justification. It is as it is.

This limit can neither be drown where one would prefer it nor can it be the result of convention. Much more, it represents the existing basis for all conventions. The basis itself is not constant and the change it undergoes is nothing else but a change of the form of life. Yet, according to Aarnio the alterations in the forms of life cannot be rationalized. They rather reflect the "limits of rationality". Thereby, Aarnio joins Wittgenstein's consideration that one passes from one form of life to another not as a result of rational arguments but rather of persuasion. The rules of rationality are valid inside the concrete forms of life but not between them.

The withdrawal from rational argumentation gets still more precarious with Aarnio's declaration that the different individuals actually comprise only a part or even a mere fragment of the form of life; thus, for example, two persons can share the same opinion about family matters or sexual morality, on the other hand, they would not necessarily arrive at a consensus on the question about the economic basis of the society.

In order to save the rationality of the theory of the forms of life, Aarnio referred to the notion of audience, introduced by Chaim Perelman, which plays a particular role in the discourse theory. Perelman himself uses 'audience' as a criterion for the distinction between argumentation and demonstration:

- The demonstration doesn't depend on the consent of any audience: Being a logical calculation it is merely correct or not correct.
- The argumentation depends on the consent of an audience, i.e. the assembly of those whom the speaker wants to influence.

According to Aarnio, this consent is a matter of the form of life. In agreement with his theory, the audience consists of "individuals who share the same form of life and who feel bound to the rules of rational discourse".

Such formulations show clearly tautology. Here it necessarily results from the absence of an explanation of the borderline notion of form of life. Although the latter remains in Aarnio's theory either vague or defined circularly, it still can be explained in the terms of a developmental theory of law, having in mind that the forms of life have not been simply given from the very beginning on as they are today – they have changed and developed.

3.1.4. *Necessity of a Deeper Meta-Theoretical Foundation of the Theory of Juristic Argumentation*

Robert Alexy himself stated that the juridical argumentation in the form of a theory of juridical discourse was open to improvement. The necessity of a further improvement, respectively of a deeper meta-theoretical foundation, however, demonstrates itself in the fact that being a theory of discourse it is very much restricted to the linguistic aspect. From the point of view of philosophy of science, the theory of juridical argumentation as a theory of discourse is still bound to the linguistic turn, and as such is to be perceived as a linguistic reduction of legal philosophy and of juridical methodology. As a beginning, this linguistic reductionism has to be abolished in the spirit of the further development (since the linguistic turn) of the modern philosophy of science. Meanwhile, philosophy of science itself has undergone some changes, which, at least partially, are quite obviously a comeback of older conceptions, preserving at that the advantages of the linguistic reflection, and, especially the advantage of formal specification. Essentially, one can summarize these changes into two mutually complementing types:

- Historical-naturalistic turn interested in the argumentation potential of evolutionary biology. It is represented by Popper and Toulmin, by evolutionary epistemology and ethics and shows an explicit turn-back to

Mach and Boltzmann. The mutual critique notwithstanding, it is related to constructivism and its genetic reconstructions of the norms of action.
- Pragmatic-technological turn, pursuing by introducing a synthetic, non-linguistic aspect the expansion, respectively, the re-forming of the predominantly linguistic philosophy of science.

Both types, which are, basically, nothing else but a change in the perspective of the classical distinction between context of discovery and context of justification, can be presented with the help of a quasi-circular or cyclic model, backed on the concept of information as a basic concept. Namely, the notion of information has a meaning beyond epistemology (See Oeser, 1976, pp. 38-40). Decision-making is founded on information, i.e. it requires and uses information. Juridical argumentation as a means of juridical ruling is in this sense a process of information processing.

Therewith the procedural character of the theory of argumentation is 'preserved' in an information-theoretical model of juristic decision. In a theory of linguistic discourse this character is stagnated to sentences as mere results of decisions and has to be constantly resolved. To count too much on a linguistic or 'rhetorical' model (Perelman) of the process of juristic decision-making would produce the impression that the jurist 'jumps' from one or more propositions, representing arguments to another proposition, representing the decision, or the legal 'ought' judgement. Thereby, the intermediate thoughts of the jurist remain lost.

Hence, the practical rationality behind the jurist's acts cannot be merely the linguistic concept and/or the screening of the used propositions in the sense of the theory of linguistic discourse. Rather, it is more probable, that the jurist would try to gather the information necessary for his decision among the already solved cases and the historically accumulated legal adjustments and on the basis of explicit propositions and implicitly pre-supposed legal and ethical principles. The adequate argumentation depends on 'pressing' as much as possible information out of a given scheme of decision-making. And although the verbal formulation of this decision is, without doubt, indispensable, it is nevertheless only a supplementary precision work, concerning the form of presentation.

Accordingly, this formulation touches only the 'surface' of the argumentation and not the arguing as a process, which remains hidden under the surface.

The essential difference between argument and proof gets lost in the linguistic turn of philosophy and philosophy of science, whereby all non-discourse and non-deductive proofs appear to be desultory and thus – almost irrational. Thereby, the concept of practical rationality remains very much obscured.

Argumentation is not only a broader and weaker term for proof – opposed to the narrower and stronger concept of deductive conclusion, but it has also a different meaning. Thus, in classical logic the argument (*argumentum probandi*) is not the proof itself, but rather the basis of the reliability of the proof, namely the grounds of proof. The latter should be differentiated from the mere form of proof or its presentation in all cases of conclusion – practical as well as with regard to the contents.

Hence, in the case of juridical argumentation, looked upon as a theory of justification, one should first of all consider the grounds of the proof and not only its (verbal) form. The next to come is the question about the origin of these grounds. The common answer to give is that they originate either from experience (observation, experiment), or from reason itself, respectively from the principles of logical thinking (axioms, definitions, verified propositions).

Accordingly, one differentiates between proofs of reason and proofs of experience. The question about the grounds of proof thus brings us to the question about the historical-genetic dimension of the theory of juridical argumentation.

3.2. *Expansion Towards Historical-Genetic Dimension: Historical Discourse*

The starting point for such an expansion of the theory of discourse rests upon the basic concepts of the theory of general practical discourse (borrowed by Alexy in the main from Habermas). For, as already shown, Habermas not only introduces the historical-genetic aspect in his elaborations, he explicitly re-formulates it into a theory of socio-cultural evolution (Habermas, 1976). To strengthen this position one has to bring in a multi-staged developmental theory of law as a necessary and adequate stipulation for the justification of the no longer disputable legal-ethical principles, the developmental processes being irreversible. Thus, the extremely weak assumptions about the social structures of consciousness (Habermas), or the ascribed to Wittgenstein formulations about cultural "forms of life" acquire a stronger argumentative character as far as one is already able to reconstruct them empirically or rationally. Something more, as "forms of life" in the literal sense, they could be even biologically explained in the terms of the genetic 'equipment' of humans. In the frames of a concept of justice on the level of the theory of argumentation, therefore, one has to take into consideration that it is not only the natural and cultural forms of life together with the legal system to undergo changes, but also the structure of the forms of argumentation. Indeed it is not possible to claim that the structure of argumentation of earlier or archaic law is identical with the methodology of Roman law, or that the latter equates with modern natural law or the German jurisprudence of concepts. Legal history teaches one primarily to methodological insights. To neglect them in favour of modern, formalistic and functioning out of history methodology would submerge rational argumentation into a drastic state of distress. The invasion of irrationality and the relativism of legal philosophy is due mainly to the ignorance in the field of legal history.

Although marginally Dreier (1981, p. 284) has explicitly stated the importance of historical discourse for a concept of justice, extrapolated to a theory of argumentation:

Let's just state how important it would be to think not only of oral but also of written discourses or discussions, and at the end – of a model of scientific progress. However in the sphere of practical knowledge this progress has not yet been too overwhelming. Yet, the historical *novum* of our present situation is given in the fact that progress begins to gain some shape in this sphere as well scientific.

As his reference to Mohr shows, Dreier advocates explicitly the "interdisciplinary discourse" between science and humanities in legal theory (Dreier, 1981, p. 284). Hence, one can conclude that, principally, he sees no hindrance for an evolutionary interpretation of the 'forms of life'. As for the historical development of legal methods Alexy (according to Raisch, 1988) has at least bound the conventional explanatory canons with the theory of argumentation. We owe the classical systematisation of canons (referred to also nowadays), to von Savigny. This systematisation includes not only grammatical-logical explanation (which according to the present view is to be equated with the logical dimension) and a systematic interpretation, but an essential and indispensable element as well, comprising the history of legal methodology. Savigny elucidated this on the basis of his reference to the historical model of methodology of Roman jurists, considered to be a typical example of legal thinking and action by all representatives of the historical school (e.g. Puchta and Ihering).

The model of the theory of argumentation has also not developed in a vacuum as Dreier himself emphasised. In this aspect, the following basic thesis can be formulated: As a contemporary discourse the general practical discourse is necessarily perceived primarily as synchronous. With the help of a diachronic, historically oriented contemplation it can be argumentatively strengthened. Alike, the special juridical discourse gets stronger on the basis of a rational reconstruction of its historical development.

This can be realized on the basis of a comprehensive presentation of the history of juridical methodology from the beginning of the civilization till nowadays. Thus, Boltzmann's statement, concerning the history of science in general can be extrapolated to legal history as well as to practical jurisprudence: The constant development of scientific method is, so to say, the skeleton carrying the progress of the whole science. To adopt this view would mean to assume that it is the development of juridical methodology, which can structure the whole development of the idea of rightful law.

3.2.1. *Roman Law and Its Paradigmatic Function in Juristic Methodology*

The importance of Roman law for juridical methodology has remained uncontested since Leibniz. Savigny, the founder of the historical school, estimates Roman law as the "model of juridical method" *per se*. In the process of characterization of this method he already anticipates the applicative circle of induction and deduction, hinting at the fact that the Roman jurists unified the theoretical and practical methods:

> Thus, their theory and practice do not really differ from one another, their theory is perfectly developed to the point of application, and their practice is ennobled on the basis of scientific treatment. In every principle they simultaneously see a case of application, in each legal case – the rule that determines the case, and one cannot but appreciate the mastership demonstrated in their aptitude to switch over from general to special, and from special to general (Savigny, 1840).

Roman law still preserves the function of a model of juridical methodology, irrespectively of the fact that with regard to the contents it is no longer valid. Ihering claims this to be the imperishable relevance of Roman law:

> Initially, being not more than a juridical grammar in the hands of those eager for knowledge, it quickly swings to the rank of law code, finally trading its widely contested and mostly lost outer authority for the much higher position of a canon of our juridical thinking (Ihering 1852-63, p. 2).

Schulz specifies this methodological problem in the contemporary literature. He differentiates between history of Roman law on the one side, and history of Roman legal science – on the other, trying to order both of them in the system of the antique history of science. According to him Roman legal science can be understood completely only in the context of the antique scientific activity. This reflects the contemporary basic idea of philosophy of science, which reconstructs the methodological structure of scientific knowledge and rational action on the basis of data from history of science. An essential prerequisite is the assumption of a continuity of methods, the constant alterations in the contents notwithstanding. In this sense one has to understand Schultz' modification of Kirchmanns' notorious dictum: "With a stroke of the pen the legislator could turn libraries into waste-paper containers. Roman legal science however remains for the eternity" (Schulz, 1961, p. 420).

Schulz classifies the history of Roman legal science into four periods:

- Archaic period,
- Hellenic period,
- Classical period,
- Bureaucratic period.

These four periods mirror the developmental periods of juridical methodology:

- The archaic period circumscribes on case law and casuistry.
- The Hellenic period marks the beginning of systematisation activity on the basis of elaboration of abstract concepts, definitions, and general legal rules.
- The classical period reaches the highest and, at the same time, turning point in the development of juridical methodology as a systematic entity of theory and practice.
- The bureaucratic period is characterized by the decline of Roman legal science, following Justinian codification. This period brings nothing new it merely stabilizes the old state.

3.2.1.1. *Case Law and Casuistry*

The archaic period of Roman legal science is defined by one general rule of legal history (See Schulz, 1934, p. 27). Legal development moves over from casuistry to abstraction. Or, to say it with other words, in the beginning legal development is characterized not by abstract general legal rules but rather by the special case (*casus*). The ruling of the particular case, then, would enter the law without any attempt at generalization. Even a better example for such an approach is presented by the legal works of the Babylonians. They supply us with the oldest known law codes of humankind. One of them, *Codex Ur-Nammu*, is traced back to the Sumerians, the second one, *Codex of Eshuna* – to the Akkadians. The best known, although by far not the oldest legal work is the *Codex Chammurabi* (around 1729 to 1687 B.C.). The *Codex of Eshuna,* composed around 1900 B.C. on two clay tablets comprises of 60 paragraphs. There one can read the following sentence: Paragraph 53, "If a bull knocks down and kill another bull, owned by somebody else, the owners have to share the worth of both bulls – the living and the dead one".

This sentence represents no general rule of law to be used in other similar cases, as for example, the principle "Bearing of the risk according to the control over the risk" (See Seidl, 1962, p. 26) does.

What is the way of proceeding in decision-making, using a law code, which is comprised solely of special case rulings and not of general legal rules like e.g. the Federal Law Code. In such a situation the judge has only one possibility – he applies the special, adopted by the law case ruling as a model for similar cases. That is to say, without any hesitation he assigns the legal provision valid (in the case of the above mentioned example) for bulls to rams. Such an assignment of a precedent to similar cases is called conclusion by analogy.

- The decision by analogy precedes with regard to the system as well as in historical and genetic aspect the decision by subsumption or deduction.

The archaic, from-case-to-case advancement predominates the Babylonian science as a whole. Parallel to Babylonian law is the situation in Babylonian astronomy. It disposes of no general principles or theories, but rather of special and concrete observations, sustained through the centuries and projected into the future. The observations, demonstrating some regularities are extrapolated to the no yet observed. Following this path, the Babylonian astronomers were able with a great certainty and exactness to make prognoses about planetary positions, and the solar and lunar eclipses, without ever possessing a theory of the solar system (See Oeser, 1979, p. 84).

Case law and casuistry, together with the respective method of analogy characterize the archaic period of Roman legal science as well. The fundamental legal work, marking the beginning of Roman law as a whole, are the twelve tablets, initially assumed to date from the middle of the 5th century A.D., nowadays dated from the 4th century A.D. The archaic period is characterized by ferocity and inexorability of law. Thus, the law permitted a creditor to imprison privately or even to kill an insolvent debtor. In some cases the creditor was allowed to cut the debtor literary to pieces. Whether, however, such acts had been really executed is highly questionable (See Liebs, 1975, p. 189). A way out of the severity and harshness of law was attempted with the help of the '*Interpretatio*'. Still, it was not used in the contemporary sense of legal interpretation, which depends on the will of the legislator. Rather, it represents a formal set of legal transactions, pursuing a goal quite different from the one of the legal provision they refer to.

The twelve tablets include the provision that a father who sells three times his son in slavery has to be deprived of control over him. The

'*Interpretatio*' reformulates the provision, proclaiming that three apparent sales liberate the son from the power of his father (*emancipatio*) (See Kunkel, 1972, p. 38).

This 'interpretative formalism' dominates mainly the sacral sphere: One may ignore a bad omen if he has not noticed it. The interpretation of the provision reads; this is valid also in the case when the person intentionally takes precautions to hinder its perception. Thus, M. Marcellus, the Roman adversary of Hannibal, had to be carried around in a covered sedan-chair each time he was awaiting a battle.

This archaic way of interpretation refers to a juridical ratio, which is still influential. Thus, nowadays, it has found a place in the German civil law code; an oral declaration of intention is to be considered void if the addressee has not heard it, even in the case when he or she intentionally covers the ears or puts down the telephone receiver (See Flume, 1975, pp. 238-240). Accordingly, the bad omen loses its power if the addressee doesn't see it – no matter whether he or she intentionally closes the eyes or covers the sedan (See Schulz, 1961, p. 35). To sum up, one can differentiate between several characteristics of the methodological structure of archaic jurisprudence:

- Case-to-case decisions by analogy, complemented by the method of formal interpretation,
- Complete lack of rational systematisation,
- Yet, already available conceptual abstractions,
- Total lack of general rules and definitions.

The conceptual abstractions do not reflect the accomplishment of a theoretical and systematic thinking. Rather they result from practice. Thus, the second oldest legal document, the *Codex of Eshuna*, offers an example for the formation of the notion of thief. "When a person sells a slave, a bull or whatever goods of value and cannot supply the name of the seller he has bought the goods from, is a thief" (See Seidl 1962, p. 29).

The archaic Roman law provides for much more complicated notions – the *manzipation* is initially meant as transfer of ownership *ex causa emptionis*. Its transformation to a transfer of ownership *ex alia causa* leads to the elaboration of the general notion of the act of transfer of ownership (See Schulz, 1961, p. 38).

3.2.1.2. The Abstraction of Juristic Concepts and General Legal Rules

The Greek influence after the Second Punic War puts an end to the archaic period of Roman legal science, or the time of the "full beards", called so only too often by Cicero (Pro Sestio 8, 19, Pro Caelio 14,33; De fin. 4). The essential change was caused by the adoption of the scientific methods, introduced by Plato and Aristotle. As Schulz correctly stated (1961, p. 81), it is this fact that transforms Roman jurisprudence into science in the narrower and more rigorous sense. As already shown, the system-character is essential for the scientific concept (See 1.3. The Structure of Scientific Knowledge).

The dialectic (Schulz, 1961, p. 81) was the preferred system-building method in the antiquity. However it differs from the narrower "antithetic" employment of the concept developed by Hegel and Marx. The antique dialectic is a broader notion, enclosing the concept of logic and, especially, the non-dialectic, linear subsumption logic. Crucial is the fact that the antique dialectic *qua* logic refers to both – the process and the result of argumentation. It paves the way "upwards" to the general principles, as well as "downwards" to the concrete reality. As Schleiermacher – Plato's German translator and commentator – correctly recognized "the initial meaning of the dialectical method" is revealed by the functional correlation between induction and deduction.

Plato offered the most comprehensive description of the dialectical method in *Politeia*, where he dealt with the problems of the ideal state. The description is based on the distinction between two types of reliable, rational knowledge – mathematical knowledge and the knowledge of general principles. Characteristic of Plato's dialectical method is the *diairesis*, i.e. he reduces an already given general notion (genus) to two subordinated concepts (species) by defining their specific differences. However the method presumes the existence of ultimate general concepts (i.e. Plato's ideas).

Aristotle opposed the a priori character of Plato's doctrine of ideas. He took account of the neglected by Plato way "upwards" and specified it on the basis of the methodological concepts of abstraction and induction.

Aristotle dealt with the way "downwards" as well. He developed the *diairetic* deduction into a demonstrative method of the "argumentation science". In the first volume of his *Analytica posteriora* he formulated precisely those stipulations, which characterized the system of the argumentative science:

- Argumentation science is a system, comprised of two classes of propositions – first principles or axioms, and deduced propositions or theorems;
- The concepts in both classes are divided, respectively, into basic concepts and deduced concepts;
- The first principles themselves cannot be proved and they are directly visible or evident;

- The basic concepts cannot be defined – they have to be directly understood, and their meaning has to demonstrate itself in the context of the system ("implicit definitions").

A text-book example for such an axiomatic-deductive system is the Euclidean geometry. However, Aristotle rejected the idea that this logical structure can be simply extrapolated to all scientific fields. Rather, the exactness of the logical proof in the particular science has to be in accordance with the corresponding subject-matter. Thus, one should not give satisfied with probabilities in mathematics, and, at the same time, one should not require logical proofs from an orator (See Stephaniz, 1970, p. 27).

The Roman jurists inherited the scientific method in this, further developed by Aristotle form. The famous "middle-man" of Greek philosophy and theory of science Cicero was able to influence the Roman jurists directly, being personally acquainted with a lot of them. Thus, he praised his friend and consul (51 B.C.) Servius Sulpicius Rufus as the first representative of scientific thought. In allusion to Plato's *Philebus*, Cicero compared him to Prometheus for introducing the divine light of dialectic in the gloomy world of jurisprudence. Somewhat exaggerating in favour of his friend and other of his contemporaries, Cicero nevertheless discerned that it was the Greek scientific ideal to bring to life the abstract legal rules – those maxims, apprehended by the classics as *regulae veterum*. This explains also the name, given to the republican jurisprudence – "regular jurisprudence". (Schulz claimed that one should better speak of dialectical jurisprudence (1961, pp. 79-80).

However, one should not overestimate the methodological consequences of the regular jurisprudence. It doesn't mean a renunciation of casuistry. Actually, it is not more than some rather simple clauses, which unfortunately do not build the system Cicero coveted in his lost manuscript *De iure civili in artem redigendo,* namely, a closed, concise system with elementary distinctions, definitions, and principles (See Schulz, 1961, pp. 83-84).

The experts in Roman law, however, wanted if not exactly the opposite (as Schulz claimed), still something different from Cicero's ideas – an "open" or "flexible" dynamical system, avoiding definitional stipulations in order to ensure a successful decision-making process in single cases. That is why the Romans chose not the axiomatic-deductive thinking but rather the topical one, oriented toward concrete problems of the singular case. The topical thinking has its roots in Aristotle and was adopted by Cicero himself. Neither Aristotle nor Cicero opposed the topic as *ars inveniendi* or heuristic to syllogistic (the theory of deductive inference); they much more considered it as subordinated to the latter. In this sense Viehweg rightly contended against Schulz and appealed for caution in opposing Cicero – as the humanists did for example (Viehweg, 1963, pp. 30-31) – as an advocate of the system approach to those jurists, who obviously neglected it. Cicero's recommendations to the jurists fit the gradual, heuristic procedure of problem-solving, which avoids abstract formulations and statements. The gap between the heuristic and topical, problem-oriented thinking and the axiomatic-deductive, argumentative, and system-oriented one is bridged by a practical discipline, used in court already at the time of the Roman republic – the rhetoric. It stimulated the birth and development of the profession of the court orator (*oratores*), which differs from the one of the educated jurists (*iuris consulti*). Servius (mentioned above) began his career as a court orator, raising himself to *iuris consulti* after receiving the necessary legal training. Cicero himself worked as a court orator, without, however, becoming a jurist in the strict sense of *iuris consulti*. Still, he developed his own rhetorical theory of legal interpretation, reconstructed by Stroux (1949) on the basis of Cicero's earlier treatise *De inventione*.

Starting point for the methodological scheme is offered first of all by those cases, in which law and its interpretation prove to be questionable. Such are the following possibilities:

- Wording of law (*scriptum*) and will of law (*voluntas*). Legal security argues in favour of retention of law. The free interpretation is backed by the consideration of *aequitas causae* or the motive of the assessed behaviour.
- Contradiction of law (antinomy or *leges contrariae*). Solution is rendered on the basis of comparison of laws, according to meaning, utility, moral value and necessity, whereby more modern law gets priority over older one, and more specific law – over a more general law. An aberration of the process of solution appears in the post-classical period of Roman law in the form of "harmonizing" interpretation, which permits the validity of both groups of laws simultaneously.
- Ambiguity of law (*amphibolia, ambiguitas*). To avoid ambiguity one should consider the language of everyday life, regard law in the context of sentences, pay attention to the origin and circumstances. As a proof or refutation one can use as well the inference from the interpretation - the so-called absurdity proof.
- Gaps in law. The syllogism is used to detect gaps.

With the adoption of the methodological structure of rhetoric the Hellenic period differentiates itself clearly from the archaic period of Roman jurisprudence, still lingering in Cato's ("truly Roman" instruction: *rem tene, verba sequuntur*) (Schulz, 1961, p. 50).

The common feature in the methodology of both – Hellenic and archaic periods, consists in the fact that the Roman jurists developed law on the basis of practice. In its own way, rhetoric protects jurisprudence from the trap of the dogmatic system.

3.2.1.3. *Systematisation and Codification of Law*

The classical period of Roman law begins with Augustine. From the point of view of methodology, it is a consequential continuation of the republican jurisprudence, bringing to the stabilization of Roman legal science. It is characterized by the end of the heuristically productive phase. At the same time, there begins the completion of Roman law to a perfectly functioning system. Cicer's sentence *Nihil est simul inventum et perfectum* is appropriately applied by Schulz in order to demonstrate the difference between the Hellenic-republican period and the classical one: "The republicans were the great inventors; the classics – only the perfectionists" (Schulz, 1961, p. 149).

Initially, the way to perfection of the legal system doesn't follow the formal-logical pattern in the sense of an axiomatic-deductive system of propositions, but rather a functional pattern. Puchta, an advocate of the so-called jurisprudence of concepts, described this pragmatic and functional refinement of Roman law in the classical period as follows:

> The goal is the complete control over the matter, the systematic penetration, the practical experience of law; the ability to trace the impact of the most general principles in their smallest ramifications and then to climb up from the most concrete case, through all intermediate levels to the very source – the first principles and at that not like in a dead mechanism but rather like in a joyous, vividly pulsating body of logically subordinated manifoldness (Puchta, 1893, p. 272).

Contemporary methodological literature still favours this depiction of the methods of Roman jurists. Max Kaser, referring to the already mentioned works of Schulz, Wieacker, etc., elaborated on the level of 'intuition', pinpointing to the casuistic character of Roman law:

> We have seen that the Roman jurists perceive the legal norms which they formulate – special norms as well as general rules – as shortened statements about experiences, gathered in the casuistry (Kaser, 1962, p. 74).

This type of Roman law has to be strictly differentiated from the system thinking of modern pandects law, which has appeared much later. One should not be misled to compare the antique Roman law with a closed system – "it has never had one and because of its classical style of thinking could have never had one". Yet, such an approach to Roman law, apprehending it only as intuitive and casuistic, contradicts the results of Horak's (1969) study of argumentation of the decision-making by the old Roman jurists. Thus, Horak demonstrated that there exist two types of methodological argumentation:

- "Conclusive" argumentation on the basis of formal logic's conclusions. These conclusions are either of a coersively-deductive type, or probability conclusions, resting upon uncertain premises, or non-cogent conclusions per analogy.
- "Causal'" argumentation, based on the empirical constraints of nature and elaborated by the addition of a special proposition of justification to the assertive proposition.

Such a meta-theory of the argumentation of juridical decision-making is necessarily directed against topical problem thinking. While Viehweg perceived the latter as characteristic for Roman jurisprudence, Horak related it to the disregard of the controversy between the method of discovery and method of justification. Dieter Nörr (1972) brought to attention a similar controversy, resulting from the extrapolation of the concepts of *divisio* and *partitio* to the contrariety between closed and open system:

- *Divisio* (Greek: *diairesis*) represents the complete list of the species of one genus,
- *Partitio* (Greek: *merismos*) represents the incomplete list of the elements of one object.

These contrarieties reflect the contemporary dispute between scientism and hermeneutic in the methodological interpretation of Roman jurisprudence. From the point of view of philosophy of science these contrarieties are not tenable because the concept of closed system itself has proved to be a non-valid fiction. And this – in a sphere where one assumes the realization of logically closed, non-contradictive, axiomatic-deductive systems of propositions, namely the sphere of mathematical theory. Using as an example the theory of elementary numbers, Gödel argued that no axiomatic-deductive system can be complete or closed in the sense that each proposition, possible for the system must be also formally deducible from the axioms. The formal systematisation provides only a relatively non-contradictive and provisionally closed system, which remains in principle open for the income of new, truer propositions, the latter being not derivable from the system. If it were not so, mathematics would have been nothing else but a colossal tautology (Oeser, 1976, pp. 77-79).

What is valid for the mathematical theories is even more valid for empirical and normative systems of propositions as, for example, the law code. Thus, with respect to the contents they are subjected to constant changes and, thereby, they necessarily represent open systems. This refers also to the codification of Roman law in the so-called post-classical or bureaucratic period, which came to an end with the Justinian codification in 534 A.C. (See Liebs, 1975, pp. 96-98, Kunkel, 1972, pp. 146-148). Methodologically seen, *Corpus juris* (called later so), doesn't form a homogenous closed system of propositions, but rather a collection of excerpts from the writings of Roman jurists. Thus, it is comprised of:

- The so-called digests or pandects dealing predominantly with private and civil law, making the greatest part of the *Corpus* (See Schulz, 1961, pp. 403-405).
- The institutions, subordinated to the pandects, which present the official introductory text-book for legal training.
- The Codex, representing a collection of the emperors' constitutions from Adrian to Justinian.

The systematisation of Roman law, completed in the form of a constrained *corpus*, takes into account all essential historical traditions. It is realized in a series of methodological steps, which could be reconstructed on the basis of the names of various jurists, quoted in the text of the digests, as for example:

Massurius Sabinus, living during the reign of Tiberius, established the fundament of the systematisation of civil law, comprised of three parts: *personae, res, actiones* (See Schulz, 1961, pp. 186-188).

Salvius Julian a contemporary of Adrian. With his 40 books *Digestorum* – a compilation of systematically arranged legal commentaries – he is the most prominent systematiser among the Roman classics and jurists.

Aemilius Papinian, killed in 212 A.D. at the order of Caracalla. Papinian refused to supply Caracalla with a legal excuse in front of the Senate for the emperor's murder of his brother.

Gaius,- a contemporary of Julian. He is the author of the only almost wholly preserved classical jurisprudence book of the institutions, comprising four volumes. Till Justinian's reign those volumes had been the standard law books for elementary legal juridical education.

Domitius Ulpian counts as the direct predecessor of the digest of Justinian I. His writings supply one third of the total scope of the pandects. Methodologically estimated, he belongs already to the deductive phase of systematic, elaborating general *regulae* to serve the purpose of deduction.

The last stage of systematisation in the works of Ulpian and his student Modestin, already shows some of the characteristics of the later defined jurisprudence of concepts. While Salvius Julian still referred to "innumerable examples" in "*ius civile* preferential treatment of the interests (*pro utilitate communi*) at the expense of logic (*contra rationem disputandi*)", Ulpian already tried to structure his contemplations strictly along the lines of logic:

- Thus he attempted to define the meaning of a word in the sense of law,
- Then he established the applicability of laws to cases,
- And, he introduced various categories of rulings to demonstrate the application of law.

An elucidating example is the analysis of the word 'to kill' (*occidere*) in Ulpian's commentaries of praetorian edicts (the first chapter of *Lex Aquilia*): At the beginning, one defines the basic meaning of the word, namely, to charge somebody "with a sword, club, or another type of weapon" or "with naked hands" (strangulation). The next step is to determine those cases, which would count as a killing. The first category of cases presents "to kill" as a direct assault against a human life: e.g., negligent killing by throwing an excessively heavy weight, a killing of a sick slave by a slight knock, or by shoving a slave over a bridge, and it is of no interest whether the killing has resulted from the knock or from drowning. The second category of cases refers to killings by an ignorant physician, or by a midwife, handing a medicine causing death – here, however, one has to consider whether the medicine has been merely delivered or personally handed, or even administered by force. In killing a slave by throwing a spear one has to take into account the particular circumstances: Has the slave been running "on the sports ground at the wrong time" or not. This refined analysis of the facts of offence is illustrated with the help of a curious example, which (although, perhaps spoiled by the activities of later compilators), provokes multifarious interpretations on behalf of legal historians:

If one of the players kicks the ball strongly, and if the ball falls on the hand of a barber, shaving a slave so that the barber cuts the throat of the former, then – so writes Mela – the ball player is to be considered guilty in the sense of Lex Aquilia. Proclus states that the barber should take the responsibility if he chooses to shave calmly at a place, (normally used as a sports ground or as a busy passage; still, it could be said – and not without right – that if one trusts a barber, whose chair is dangerously placed, one should blame nobody but himself) (Schulz, 1934, p. 43).

A third category comprises those cases where more than one person participate in the killing: e.g., one holds the slave, while another kills him, or several persons either at the same time or one after the other beat the slave.

A culmination in concept abstraction offers Ulpian's popular definition of natural law:

Ius naturale is what all beings have been taught by nature: This law is not an exclusive possession of human race, but rather of all beings born on the Earth, in the sea, and of the birds, as well. Hence, the relation between man and woman comes which we call marriage, the begetting of children, their rearing. We see, namely, that also the rest of the beings, including the wild ones, will be judged on the basis of the legal experience (After Seidl, 1962, p. 100).

Ulpian and his contemporary Paul are the most quoted Digests' jurists. Already before Justinian, the post-classical tutors of law, borrowing *norma* and *regulae* from classical jurisprudence, praised them as the coryphaei, as the perfectionists of the latter. The tendency to see classical systematisation as a unsurpassable master-piece (*Akme*) gives also the impetus to compile the works of leading authors into a closed canon.

The first attempt realized in this direction before Justinian's codification is the so-called law of citations by Theodosius and Valentinian in 426 A.D., which in 438 A.D. merged with the *Codex Theodosianus*. Most probably, the initial version restricts the citations to the works of Papinian, Paul, Gaius, Ulpian, and Modestin. According to the version adopted in *Codex Theodosianus*, also those jurists, mentioned by one of the five above, acquired the right to declare law with authority. However, this later stipulation caused the failure of the idea to form a closed canon – alone Ulpian quoted other authors in abundance (See Schulz, 1961, pp. 357-359). Yet this could not stop the tendency which began with the systematisation of Roman law by particular authors and which was practically absent in the period of the Republic and of the classical principat: The tendency towards stabilization of law by giving the legal rules a state status, thus withdrawing them from further juristic disputes.[3] The process begins during the reign of Diocletain, and it is traced in all imperial constitutions (*Codex Gregorianus* and *Hermogenianus*), to reach its climax at the time of Justinian. According to Schulz (1961, pp. 365-366), the Digests are "the most excellent and comprehensive stabilization of Roman legal system". The legal rule is to be already conceived as an "emperor's resolution". In *Constitutio Tanta*, publishing the Digests, one reads: "All dicta of the classics, gathered in the Digests, are to be treated as if written by Justinian himself or with his personal permission."[4]

Herewith, Roman law reaches its *competens forma* (*Constitutio Tanta pr*), which, from the point of view of methodology, is first of all a terminological unification on the basis of interpolation. The method of interpolation, used not only by Justinian's legal scholars, but also by post-

[4] See Const. Tanta § 10.

classical jurists, is, by itself a fundamental accomplishment in abstraction. Thus, *interpolatio* means, primarily, the establishment of a more general link between two known notions. Hence, the transformation of texts on the basis of interpolation represents, in the first place, the substitution of a special notion by a more general one (e.g. *mancipatio* by *traditio*). This fixation of the definitions of particular notions is an additional evidence for the stabilization of the legal system with the help of codification.

3.2.1.4. *Transformation and Further Development of Roman Law in the Middle Ages*

The continuance, interruption and recommencement of Roman law import in the early Middle Ages is mostly beyond knowledge, due to the scanty and doubtful resources. That little from Roman jurisprudence which survives the political and intellectual collapse of the old world, reaches us in a vulgarised form and incomprehensive style. The question whether Roman legal education was further carried on at the two focal points of Byzantine culture – Rome and Ravenna – is disputable and dealt with within the so-called "continuity theory". However, it is a fact, that the rediscovery of Roman law in the form of Justinian's codification is traced back to the university of Bologna (See Koschaker, 1958, pp. 55-57, Schlosser, 1979, pp. 17-19). The revival of Roman law has to be interpreted in the comprehensive, historical and scientific context of mediaeval attitude towards antiquity. Antique science, particularly, antique philosophy is subordinated to scientific reason. The organ of reason is logic, which attains solely to ascertain itself in the incontestable truth of authority. According to Wieacker (1967, pp. 6-7), this relation between authority and logos determines not only mediaeval philosophy and theology from the perspective of the "scholastic method", but jurisprudence as well:

- As a *ratio scripta* the single juridical text (*locus*) is a truth for itself, the context of all texts together being of no importance. One has to ascertain oneself of the meaning of the single text, determining the basic form of the current explanation of the text – the annotation (*glossae*).
- The method of application is based on the application of the Aristotelian-scholastic syllogistic and of Aristotelian-scholastic basic concepts like *causa formalis*, *materialis*, *effeciens* und *finalis*, resulting from the subdivision of the concept of cause, and including the technique of dispute. Of particular importance is the adoption of the Aristotelian theory of definition, based on the concepts of *genus* and *species*, as well as the adoption of the typical scholastic procedures of *distinctio*, *divisio* and *subdivisio*.

- The glossators do not limit themselves to the consecutive explanation (exegesis) of particular places in the text. Being convinced of the rational line, governing the written heritage, they attain to fathom the reason in the texts as a whole, and to establish formally non-contradictive syllogisms. Thus, they employ a harmonious exegesis, trying to reveal the compatibility of contradictory texts. Continuing these activities, stimulated already by Justinian himself, the glossators benefit from the almost inexhaustible wealth of juridical texts, their inter-changeability, and possibilities for generalization.

The harmonization efforts of the glossators are later differently evaluated. Thus, Schulz' attitude is clearly negative:

These distinctions and resolutions of contradictions to a big extent do not possess juristic value, being not the result of juristic reflections about juristic problems, but rather the result of an ivory-towered pseudo-philology (Schulz, 1961, p. 419).

Yet Wieacker stated differently:

It is only with the help of an unrelenting comparative pervasion of the sources that the glossators could adopt all the juristic problems of *Corpus Iuris*. By exegesis, harmonization and formation of rules, they established a didactics of (formal) non-contradictive propositions – perhaps, the first altogether autonomous, juristic dogmatic in the world's history (Wieacker, 1967, p. 59).

From methodological point of view, these different evaluations of glossators' activities reflect the double-sidedness of each systematisation process: The more general the basic principles of a system are and the more precise the binding of deduced propositions into an entity is, the more removed is the system from the concrete reality of the particular cases. Thus, the latter could be achieved only on the basis of long deductive processes, and if the deductive processes stick into general formalizations – this reality cannot be achieved at all.

However the realization of the gap existing between the perceived as timeless validity of Roman law, on the one hand, and the historical character of the reality, on the other becomes the essential impetus for a methodological change. This gap could be bridged only by the fiction that the validity of Roman law is indirectly derived from the identity of the concept of authority. Hence, in Italy as well as in France, one had to learn juristic "grammar" and not the adequate for the particular sphere law (See Wieacker, 1967, pp. 80-81; Vinogradoff, 1929, Koschaker, 1958).

Roman law and legal reality were brought together thanks to the efforts of the so-called "post-glossators", "commentators", "practitioners" or "*consiliatoren"* (consultants). While preserving the methods on theoretical level, they, in addition, considered the current practical problems, creatively structuring Roman law in free commentaries. Therewith, they closed the methodological circle of case, hypothesis of laws, system of axioms and practical decision, the latter serving as a constant corrective and a basis for the further development of the theoretically stabilized system.

3.2.2. *Juristic Methodologies of Modern Times*

At the beginning, the modern times' development of the juristic method is still closely linked to Roman law. In mediaeval Europe, the latter is looked upon as the original legal system, or as the natural law in the virtue of spiritual authority (Wieacker, 1967, p. 145). Roman law, as an imperial and Reich's law, was a representative of legal uniformity. Thus, in the case of lack of rules in a particular legislation, e.g. municipal legislation, Roman law was subsidiary valid. Since 1495, it had been applied by the Supreme Chamber Court of the German Reich (See Koschaker, 1958, p. 90, pp. 118-120). A methodologically important and relevant change took place, however, already in early modern times with the so-called *Usus modernus*. Originally, the notion "*modernus*" referred to the activities of the glossators and commentators as opposed to the antique *Corpus juris*. In the 17th century, it was being used already to dissociate one from the famous mediaeval authorities as a whole. *Usus modernus pandectarum*, borrowed from Samuel Stryk's (1640 – 1710) work of the same name, meant as well a *Nova practica*, a "new, time-relevant practice of Roman law", which emphasized the new attitude to Roman tradition, referring no longer to its theoretical reception. The complete victory over this theoretical tendency in Germany is linked with the name of the politician and jurist Hermann Conring (1606 – 1681). The more liberal relation to the Roman sources, the validity of which was considered no longer automatically as irrefutable, proved to be inherent for the pragmatic jurisprudence. In its turn, this strengthened the autonomous legal awareness as well. Yet, the territorial differentiation of legal systems had also external political reasons. Thus, the thirty years war witnessed the collapse not only of the universal way of thinking of the Reich, but of the universal legal awareness, legitimising Roman law as well.

For pragmatic reasons, *Usus modernus* broke with the existing juridical methods, which were based solely on authority and reason. At the same time, it left behind a kind of theoretic-methodological "calamity" (Wieacker, 1967, p. 251), which could not be legally settled by the humanism, developing in the 15th and 16th centuries. Thus, for example, in improving the methods of text interpretation, the humanists only replaced mediaeval authorities by antique ideals. In the long discussion, concerning the correct methods, the humanists focussed not on the juridical cognitive and decision-making methods, but rather on the educational. It is true, that some attempts at axiomatisation of law could be traced back to that time. Yet, compared to the standards of the antiquity, those attempts are quite simple and strike one as odd. An early example of this type offers the educated also in mathematics jurist Sebastian Derrer, teaching jurisprudence in Freiburg in the first half of the 16th century. In 1540 Derrer wrote *Jurisprudentiae liber primus, instar disciplinae institutes et Axiomatibus magna ex parte conscriptus* (Freiburg i. Br. u. Lyon), where he used as "axioms" literally quoted sentences by Roman jurists, particularly by Ulpian. The proceedings demonstrate clearly that these and other similarly executed arbitrary and artificial attempts at systematisation obviously lack a true foundation.

The legal thinking of modern times was based on the renewal of the old idea of natural law, methodologically supported by the cognitive ideal of the modern natural sciences (See Welzel, 1955, Koschaker 1958, pp. 245-247). This development was generated inside the natural sciences themselves. Thus, for example, in his cosmic harmony, Kepler extended the mathematical and geometric proportionality to human society, while Newton drew a parallel between the history of development of physical Universe and its laws, on the one hand, and human history, on the other (See Oeser, 1975).

Hence, the application of the methodological ideal of modern natural sciences in jurisprudence is not just a supplementary expansion and transfer it has rather been present from the beginning on. The natural-scientific method in its essence is not the invention of modern times; it reflects the antique idea of scientific knowledge acquisition, comprised necessarily of induction and deduction (See e.g. Welzel, 1958, pp. 43-45).

In describing the *metodo risolutivo*, Galileo (1891, p. 220) referred explicitly to Aristotle whose scientific system is deductively built but whose reflections in the process of the system's formation could have been only inductive. The new in this method, based on opposite operations:

- Metodo risolutivo – metodo compositivo (Galileo), and
- Analysis – synthesis (Newton),

is abandonment of the knowledge of the essential, imprinting the cognitive ideal of antiquity. To a certain extent Plato's pyramid of concepts lacks the apex, where the first ideas (Plato) or, the causes (Aristotle) are housed. In his letters about the sun spots, Galileo proclaimed his abandonment of the knowledge of the essential by writing that one either tries with the help of speculations to penetrate the immanent essence of natural substances, or one has to content himself with the knowledge about some of their empirical features ("*affezioni*"). In his *Scholium generale* Newton stated that while he had not been able to discover the cause of the properties of gravity. However it was enough that gravity did really exist, and acted according to the explained laws, and abundantly served to account for all the motions of the celestial bodies, and of seas (Newton, 1872, p. 511). The possibility to verify empirically and to correct the axioms of empirical scientific theories is a methodological consequence of the abandonment of the search for absolute principles. Therewith, the methods of induction and deduction were functionally linked, which proved to be of crucial importance for the further development of natural science. The modern times' methodological awareness of jurisprudence became part of this development. The link between both fields – natural sciences and jurisprudence – was not belatedly supplemented it had rather existed already in the idea of a unified system of scientific reason ever since the beginning of modern natural sciences.

The methodological link between these two, with regard to the contents completely different fields, is responsible for the lasting till nowadays in jurisprudence methodological debate between hermeneutics and scientistics. Before concentrating on this link, however, a very popular misunderstanding, concerning the notion of closed system has to be cleared.

Legal historians often claim that the axiomatic-deductive, closed system is the logical form accomplished by natural sciences – if not yet by Copernicus and Kepler then by Newton in his *Mathematical Principles of Natural Philosophy.* Thus Wieacker claimed:

> The reduction to relations, which can be expressed in numbers is the prerequisite for the formation of generally valid natural laws, because it is the measurements that permits the comparison of unalterable recurrent processes... The experiment with nature aims at the ascertainment of special natural laws, and – on this basis – the derivation of more general laws, and axioms. The closed systems of that time were built in this way, namely, on the basis of such an ascent first, to the more general, and then – to the most general: e.g., Newton's physical world picture (*Philosophia naturalis principia mathematica*, 1687), Spinoza's *Ethica more geometrico demonstrata*, and last but not least – the system of law of reason.... The modern claim to know natural law is now extrapolated to the nature of society, i.e. to law and state. The latter have to be provided as well with "ought" laws on the basis of the unchangeable certainty of mathematical inferences. The logical correlations of the natural laws form the system of the physical world, culminating in Newton's *Principia mathematica*. In the same way, the natural laws of the social world form a closed system of society, a "natural law" (Wieacker, 1967, pp. 254-256).

This characterizes the stand-point of modern hermeneutics in the sense of Gadamer who sees the realization of the scientistic programme already in the beginnings of modern times' natural sciences.

The historical, and thus hermeneutical analysis of original texts shows that, on the contrary (see Oeser, 1979, pp. 53-55), even in the case of Newton one cannot speak about a closed system, based on logical deduction: no concept is so completely defined in the sense of mathematical-logical exactness that it could not be further altered, and no theorem is so self-evident that it would not require a concrete proof of its deductibility from the axioms.

The classical theories of terrestrial and celestial mechanics with their cumbersome mathematical apparatus of Euclidean geometry are far away from being logically non-contradictive deductive systems of propositions. That's why neither Galileo nor Newton spoke about deduction. It is the functional counterpart of analysis, particularly described as induction, which represents the constructive synthesis. As a system-building method however, the constructive- synthetic method comes before the axiomatic-deductive system itself. As Newton formulates it, this method consists solely in the "acceptance" of the inductively acquired principles.

The "ascend" to the most general laws or axioms should not be understood as a derivation in the sense of logically correct inferences. Rather, it has to be seen as inductive generalization, realized step-by-step by inductive analysis of the multifarious conditions of an event, which leads at the end to simple basic laws.

Hence the act of inductive, supported by observation and experiment construction of the axiomatic theory basically represents a process of constant adaptation between the more abstract theory, on the one side, and the empirical experience, on the other, which doesn't permit a specification of definition of the basic concepts.

The most popular example for this is the ambiguity of the concept of mass in Newton's work: "inert mass" in the axioms, "heavy mass" in the law of gravitation. However the latter was propagated by Newton as deducible theorem according to the logical structure of his system.

The laws of Galileo and Kepler reveal a similar situation. As sub-theorems they should be deducible from the law of gravitation, but in fact they are just adapted to the system by forced reconstruction. Also, the solution of the concrete and single problems of explanation and prognosis of particular events in nature (e. g., planetary motion, high and low tide, etc.) is not to be regarded as a simple process of deduction and (mathematical) calculation, but rather as an adaptive process of construction including many heuristic elements.

Therefore, methodologically seen, classical constructive and synthetic theories of science resemble the "open system" of the "casuistic and pragmatic" Roman jurisprudence much more than it is usually seen in the literature on legal history. Neither Galileo's, nor Newton's systems are dominated by logical consistency; their consistency is a rather pragmatic one.

Axiom and observation do not form a "dualism" (Wieacker, 1967, p. 254), but represent a functional unit. The dualism or "opposition" of induction and deduction has its roots in extreme philosophical positions in modern epistemology – i.e.Baconand Descartes on the other. Galileo had been attacked by both. Bacon imputed to him (and the Italian astronomers) an unwarranted mathematical anticipation of nature that can be grasped only by inductive experience. Descartes, on the contrary, criticized Galileo on the basis of the assertion that his *modus operandi* was inductive and failed to recognize the principles. In this methodologically somehow confused situation, the systems of natural law (Hobbes, Spinoza, and Pufendorf,) were born not as mere "copies" (Stephanitz, 1970, p. 65) of the scientific and mathematical systems of geometry and mechanics, but as autonomous achievements.

3.2.2.1. *The Beginnings of Methodological Reflection in Modern Times Theory of Natural Law*

Ever since antiquity, the idea of natural law can be found in all cultures and societies. Basically, it includes a mere theoretical and philosophical claim. But opposed to positive law, it has always had a critical function, because it has challenged legal reality, which depends on historical constraints, political power, interest, and social structures. The theoretical claim can be put into practice when it is linked with irritated social groups – with people irritated by injustice and power. In such a situation, the traditional conception of law can collapse, and people are looking for a new legitimisation of law (Wieacker, 1967, p. 257).

In modern times, the efficiency of the doctrine of natural law has been due to rational scientific method. Unlike traditional antique and medieval conceptions, natural law has been "profane law of reason". Its revolutionary effect has been a "total, pervasive ideologisation" of public life. The *Corpus iuris* had shown the European jurist how to discover the specific problems of any single legal case. As a rational natural law, however, reasonable law determined public action at the beginning of modern times, no matter to whom or to what the action was related: to the purpose of the sovereign and the nation, to religious groups, to social classes, or to the individual (Wieacker, 1967, p. 272). But like any change, this "mental revolution" was anticipated by a closed tradition, which can be traced through the social doctrines of the Middle Ages back to ancient philosophy. According to the studies by Troeltsch (1925, pp. 166-168), Wolf (1964), and Welzel (1955) but also by Schlosser (1979, pp. 38-40) (which were followed by Wieacker), this continuity of the development was determined by the unity of problems and can be brought into the following system:

- *Ancient natural law*:Natural Law had been introtruced by the sophists. They trie to establish (partly realistic and, partly, naturalistic) the lewfulness of human society. It appeared in Plato's doctrine of the ideas as generally valid and necessary knowledge about justice, and was systematically arranged and filled with content by Aristotle. Finally, natural law developed as a cosmopolitan idea in stoic philosophy (Chrysipp, Zeno) and was conveyed by Cicero to Roman jurisprudence (where, however, it did not gain any direct influence).
- *Christian natural law*followed the ancient idea of justice, but put natural law into relation to the immense justice of God. It found its most clear expression in the differentiation between *Civitas Dei* and *Civitas terrena.* According to Augustine this differentiation does not simply mean two separate ontological levels but a dividing line that runs through all kingdoms of the world, which without God's true justice would be nothing else but a "barbarian riot". Only with Thomas Aquinas, however, Christian natural law developed as a rationally founded legal metaphysics, which included the predominant role of reason. On the contrary, Duns Scotus and particularly William Ockham established the idea of Divine justice – God's free will, so to speak – which cannot be reduced to rationally understandable world principles, but are inscrutable decisions of God concerning the obligation of humans. Yet all these different conceptions of Christian natural law have in common the belief that *human* natural law is subordinated to the revealed Divine law that gives the ultimate standards.
- *Modern natural law*, due to reformation, appeared first in different faces. The Spanish jurists (Fernando Vasquez, Covarruvias y Leyva) and moral theologians (Gabriel Vasquez, Francisco Suarez) stuck to the Thomistic tradition and tried to combine their personal experiences (war, trade and economy) with Christian natural law by means of the method of "middle principles". The advocates of reformation took a different view. Luther considered the profane natural law as a temporary principle of order, just to avoid total corruption. Calvin who was a jurist took the same law as a means to achieve – in the struggle between reformation and counter-reformation – the very Divine kingdom. This idea was seized by Johannes Althusius, the eminent founder of the modern doctrine of people's sovereignty, who had been inspired by the Calvinist concept of community (See Wolf, 1951, pp. 176-178).

However, it is said that the very founder of modern natural law was Hugo Grotius (1583 – 1645). He came to this reputation because of his famous work *De iure belli ac pacis libri tres* (1623), which was devoted to international law beyond religious belief and more or less independent of moral theology. Grotius correctly stated that this task can be done only on the basis of the insight that there is natural law which unifies all human beings.

His method was different from both the methods of his theologically determined forerunners, and the methods of his successors who were oriented toward the modern conception of science. He used all available evidence of legal experience in human history – from Roman law via medieval theology to humanism of modern times – and was a pioneer of comparative law. Nevertheless, his very ambition was not comparative jurisprudence, which he used only to establish a generally valid natural law. In contrast to the *ius divinum*, such a law, he claimed, would be valid even in the case that God does not exist. Also, contrary to the *ius civile* (*inter arma silent leges*), it would have its function in times of war, too.

The comparative method was considered by Grotius as an empirical evidence for the universality of certain legal principles and institutions. However, he did not escape from "hypostasis", i.e. the arbitrary overestimation of empirical and historical legal principles and their interpretation as principles of natural law. Hence, as Wieacker correctly stated, Grotius did not yet succeed to establish a methodological innovation of positive legal science in the spirit of natural law. This step was possible only on the basis of the philosophical achievements by Hobbes and Spinoza who applied modern scientific method to the sphere of human society and ethics.

Hobbes published his natural-law doctrine first in his *De cive* in 1642 and later in his *Leviathan.* He was a contemporary of Descartes and Galileo, whom he probably visited in Florence in 1636. His methodology is connected with the Cartesian method in a somewhat peculiar way. In the preface to *De cive*, Hobbes characterized his method as follows:

Concerning my Method, I thought it not sufficient to use a plain and evident style in what I have to deliver, except I took my beginning from the very matter of civil government, and thence proceeded to its generation, and form, and the first beginning of justice; for every thing is best understood by its constitutive causes; for as in a watch, or some such small engine, the matter, figure, and motion of the wheels, cannot well be known, except it be taken in sunder, and viewed in parts; so to make a more curious search into the rights of States, and duties of Subjects, it is necessary, (I say not to take them in sunder, but yet that) they be so considered, as if they were dissolved, (i.e.) that wee rightly understand what the quality of humane nature is, in what matters it is, in what not fit to make **upa** civil government, and how men must be agreed among themselves, that intent to grow up into a well-grounded State. Having therefore followed this kind of Method (Hobbes, 1983, p. 32).

These words have been understood in connection with the at that time quite popular metaphors of man as machine and have been criticized as rather uncritical application of mechanistic thinking to human society.

According to a common conviction in legal-historical literature, Hobbes marked the beginning of the strictly logical deduction of secularised natural law from "empirical principles". Therefore, he has been considered as the protagonist of the naturalistic fallacy, which has accompanied the history of natural law for centuries. This fallacy is known as deduction of *ought* from *is*. Hobbes was in fact aware of the fundamental difference between nature and human society. According to him one could not explain the characteristics of natural objects from their causes. Rather one could infer these characteristics "backwards" - from those features that are visible to one. Politics and ethics differ – there it is we ourselves who create the principles of justice and injustice (Hobbes, 1949, p. 17).

The bridge between natural law rooted in the essence of human nature, and positive law is a peculiar construction. It is the model of the social contract.

The social contract does not mean a single event in early human history, but rather indicates a timeless and "super-historical" truth. It is indeed represented in a narrative way however, it serves as model for the consistent foundation of law.

The idea can already be found in ancient natural law (Plato, Aristotle, Cicero, the Stoics), but only in modern times it has gained its fundamental meaning as a methodological construction (Suarez,., Althusius, Hobbes, Spinoza, Lock, Pufendorf, and Rousseau).

The modern version of the social contract refers to the human individual in a primordial state which, again, does not mean a specific (historical) state in the development of humankind but the "natural state" or "essence" of humans. Hobbes did not attain at this definition by empirical evidence in the sense of (empirical) observation in the natural sciences. He rather attempted a general truth, based on the inner self-experience of man:

> In the first place I set down for a Principle by experience known to all men, and denied by none, to wit, that the dispositions of men are naturally such, that except they be restrained through feare of some coercive power, every man will distrust and dread each other, and as by natural right he may, so by Necessity he will be forced to make use of the strength **hee** hath, toward the preservation of himself (Hobbes, 1983, p. 32).

This "elementary principles" received its vigorous expression in the famous formula *homo homini lupus*: Each human being is a wolf for the other humans. Hobbes was aware that this sentence can not be the axiomatic fundament of a theory of justice. He only wanted to characterize the human state of nature, before there was any contract or law. The question, whether humans according to their very nature are "good" or "bad" was even not decisive because in that state one could hardly make the difference. Thus, Hobbes maintained that before there was agreement and law, humans – like animals – were not aware of justice or injustice and did not have any notion of "good" or "evil". Before the emergence of civil society, he said, there was a war of all against all, for any individual desired to preserve its own right, to overcome his miserable state, it was necessary to establish contracts.

Hobbes saw the "first and elementary law of nature" in the positive norm that one is looking for peace and, when one cannot find it, to retreat to war. How to preserve the "inner peace" in a human society? According to Hobbes the association of equal people is not enough; in order to guarantee security, all people have to subject to a "person" which is the state and which can be represented by an emperor or an assembly.

Hence, Hobbes' structure of argument is not determined by an axiomatic-deductive demonstration of logical inference, but depends on the frequently discussed practical inference that gives specific normative rules and so helps to reach the "axiomatically" presupposed goal. The same method he applied to the presentation of his basic statements concerning declaration of intention and commitment to the contract:

> But the act of two, or more, mutually *conveighing* their Rights, is call'd a *Contract*. But in every Contract, either both parties instantly performe what they contract for, insomuch as there is no trust had from either to other; or the one performs, the other is trusted, or neither performe. Where both parties performe presently, there the Contract is ended, as soon as 'tis performed; but where there is credit given either to one, or both, there the party trusted promiseth after-performance; and this kind of promise is called a Covenant (Hobbes, 1983, p. 55).

Pufendorf (1632-1694) grasped these ideas and developed them further. (See Wolf, 1951, pp. 306-308, Welzel 1958) He got to know the doctrines of Grotius and Hobbes through his teacher – the mathematician and philosopher Erhard Weigel (1625-1699) in Jena, and presented his system of legal theory in the *Elementorum jurisprudentiae universalis libri duo* (1660) and in his textbook *De jure naturae et gentium libri octo* (1672). In the *Elements* the axiomatic structure is given in the following way:

- Two "rational" principles, which are called "axioms" and express the idea that humans are capable of responsibility and able to bind themselves to agreements. These principles (as mere rational principles) come directly from reason.
- Five empirical or "experimental" principles which are called *observationes* and show that humans are capable of judgement, free will, self-love and a social instinct and are weak in nature. The certainty

expressed in these sentences is based on observations and comparative experience.

Pufendorf concluded that every person in his/her aspiration for self-preservation should act only in such a way that human society is not threatened, for otherwise he/she would act against his/her own interest (See Stephanitz, 1970, p. 69). This problem he elaborated, with some insight, in his *opus magnum* – the eight books on natural and international law. Although the link between axioms and observations is methodologically relevant it doesn't equal the methodology of natural sciences. The rational axioms are not scientific law-hypotheses, but ethical axioms that give natural law the character of a "social doctrine of obligation", and the observations are not observations of physical states (*entia physica*), but refer to objects of the social world of moral freedom (*entia moralia*).

The functional connection between deduction and induction – Pufendorf's ambition in his system – is therefore not located at the level of theoretical knowledge and empirical evidence but at the level of moral and normative claims and the concrete action within legal reality.

This idea was preserved by Leibniz who underlined the normative character of law as based on will (*voluntas*), and not on reason (*ratio*). Yet Leibniz strictly advocated the idea of a rigorous mathematical demonstration of natural and positive law: According to him jurisprudence is one of those disciplines that do not depend on experiments and sensual experience but rely on definitions and rational evidence. Justice consists of a certain congruence and proportionality (*congruitate ac proportionalitate*), so that something can be recognized as just, even if there is nobody who experiences justice. In the same way the power (ratio) of numbers does not depend on a person who is counting and on objects, which can be counted. Also, our sense-organs are of no relevance in this context. What is important, however, is a clear imagination, which was called "idea" by Plato and which, expressed in words, is the same what we call "definition" (After Stephanitz, 1970, p. 75).

Leibniz himself did not realize this plan. The latter was given a concrete shape by Leibniz's disciple Christian Wolff in his *Ius Naturae methodo scientifica pertractatum* (1740-1748). Comprising 8 volumes, this is the most complete work on natural law.

However, the realization of the "demonstrative" method did not follow Leibniz' ideas. Somewhat later Leibniz replaced deduction by combinatorial operations which he saw as the basis of juristic technique. Thus, the legal system should not be developed by the method of deductive separation of concepts, but by combining elementary juristic notions. That is why for

Viehweg the application of the *ars combinatoria* to jurisprudence is a mathematisation of topics. After many useless efforts, Leibniz himself abandoned the thought that moral ideas could be represented by elements and their combination, and was content with the possibility that definitions are of good use in all moral matters and can help to avoid difficulties by recognition.

However Christian Wolff advocated an earlier view developed by Leibniz in his younger years in Nova Methodus. With admirable consistency Wolf took to the deductive systematisation of natural law belonging – according to the understanding – to philosophy, which could be deductively demonstrated as well. Actually Wolff went beyond it by formulating the effective thesis about the demonstrative character of (Roman) positive law. As he argues in his paper *De civili iurisprudentia in formam demonstrativam redigenda*, the demonstrative character of positive (Roman) law is to be explained with the fact that firstly it is derived from natural law (the latter having anyway demonstrative character) and secondly, that as Corpus Juris according to the wish of Justinian it should consistently seize all legal cases. The system gains positive character through its inversion. The legal material resolves in clear sentences, which allot to a subject (or an object) something concrete (the legal processing). Afterwards one should look for the grounds behind the sentences and should formulate them as principles. Then one should derive through abstraction possibly fertile definitions from these sentences and principles. At last one should arrange the whole into an easy to survey logical system. Inverting the order, one could infer legal sentences by chain-syllogisms from the definitions and principles. (After Stephaniz, 1970, pp. 88-90).

With his thesis Wolff not only stimulated a series of demonstrative presentations of positive law but he created the foundation of juridical dogmatic of the 19th century. Juridical construction gained its importance as a logical and definitional concept formation in this very dogmatic.

3.2.2.2. *Historical School*

The historical school, founded by Savigny, could not be understood without an account of Kant's influence. Thus in his philosophical works, and particularly in the *Foundations of the Metaphysics of Morals*, *Critique of Practical Reason*, and *Philosophy of Law* Kant replaced the non-critical and based on absolute and determined axioms natural law with the principle of moral autonomy of the person.

This autonomy is best presented by the categorical imperative: "Act only on that maxim through which you can at the same time will that it should become a universal law" (*Kants Werke,* 1968, p. 421). According to Kant on this basis one could define law independently of any content-reference to the essence of man. In this spirit law represents the conditions for the unification of the arbitrariness of one with the arbitrariness of the others according to a general law of freedom (Kants Werke, 1968, p. 230).

On the basis of this general definition of law and through the simple consideration that the authority of obligation is necessarily (e.g. "according to the principle of contradiction") connected with a law founded on the notion of freedom Kant defined "strict law". The latter could be presented as the possibility for freedom of each and all according to general laws of a harmonized universal and mutual coercion (*Kants Werke,* 1968, p. 232). This idea of strict law underlines Kant's systematic doctrine of law, which he calls *Metaphysische Anfangsgründen der Rechtslehre* (*Metaphysical Foundations of the Doctrine of Law)* as an analogue to his *Metaphysische Anfangsgründen der Naturwissenschaft* (*Metaphysical Foundations of Natural Science*). In a methodological aspect this analogy is no way a non-binding metaphor. Rather it determines totally the structure of the doctrine of law. Thus in the introduction Kant compared the construction of the concept of common freedom with the possibility of free movements of the body based on the law of equality between effect and counter-effect. According to Kant the pure reason determines one's due with mathematical precision in legal science as well (Kants Werke, 1968, p. 233).

Kant looked for the sources of the true system of law in the pure reason. For him the empirical doctrine of law is nothing but a "wooden head" without a brain. Savigny on his part drew other conclusions from the basic principle of the "free existence of the individual" and the necessary "mutual restriction of the will". According to him the degree of restriction of the individual could not be dependent on the will either of another person, or of something else, i.e. the state. That is why there should be something that is totally objective and independent, wholly aside from any individual belief – namely the law (Savigny, 1951, p. 14, Larenz, 1975, pp. 11-13).

According to Savigny the "objectivity" of law consists in the fact that the person who applies it "doesn't need to add to it anything himself". In the tradition of the linguistic usage of the scientific methodology of his time Savigny called the whole knowledge derived from what is objectively given a "historical knowledge". From this he concluded that the "whole character of legislation science should be historical":

- Historical in the literal sense of the word, e.g. laws are conceived as objective facts in a "time-series";
- Philological, e.g. laws are linguistically formulated.

The utter objectivity of law and its existence as a linguistic entity brings Savigny to the momentous conclusion that the job of the judge who applies it consists in its "pure logical interpretation". Such a logical interpretation of laws is however possible only if law functions as a system. That's why legislation science should be not only empirical and historical but also a systematic one. In this sense jurisprudence abuts up on philosophy, which "has to present the whole extent of the more general proposition through a complete deduction".

Both Kant and Savigny founded law on the basis of the moral autonomy of the individual. The difference between them consists not in the controversy between historic-philological and systematic-philosophical methods but rather in the shift of focus: Kant gave priority to the systematic-deductive method. The historical method is secondary – it provides the deductive system of pure reason of law with contents. In the case of Savigny the historical method has a priority. Nevertheless it is from the beginning on linked to the systematic method.

As a "common" "historical" revision of jurisprudence the historical method would lead to an "external history of law" which as a part of political history tells the story of the intended changes in law. However as a "historical in the true sense of the word" revision of jurisprudence this method leads to an "internal history of law" which accounts not only for particular legal problems but rather conceives law as a "constantly developing system" thus revealing "the interdependencies in the whole". The common, e.g. the "narrative" historical method thus turns into a "history-of-development" method, which links system and history into a functional entity.

With programmatic brevity and clarity Savigny formulated the basic structure of his juristic methodology in the introduction (recorded by Jacob Grimm) of a course of lectures hold in the winter term of 1802/03. He claimed that to set up the system historically according to time and place one had first to resolve the legislation into its single elements and to present the latter in their true correlation (Savigny, 1951, p. 17). One has also to take into consideration the perfect structural identity of this model of juristic methodology with

- The antique 'dialectical method' based upon induction and deduction;
- The natural-scientific methods of Galileo and Newton of modern times (*metodo risolutivo* and *metodo compositivo,* analysis and synthesis);
- The propagated by Pufendorf (in the tradition of natural-scientific method) method of natural law (observation-axiom).

Savigny's model consists of three parts: philological, historical, and systematic-philosophical one. In all parts he offered important specifications of basic methodological notions: Thus, he understood the notion of juristic interpretation as a "contents-reconstruction of law" e.g. the interpreter has to embrace the stand-point of the legislator in order to reveal his claim. One has to differentiate between juristic interpretation and the common notion of interpretation, the latter being responsible only for the clarification of obscurities. Juristic interpretation should account for three aspects:

- Logical aspect, which for Savigny consists not in a formal-logical presentation, but rather in a genetic contents-presentation of law. The thought-contents of law has to be orally pronounced, hence the necessity of the rules of speech. This leads thus to the
- Grammatical aspect of interpretation, which should be complemented by the
- Historical one. Namely, law is given to a particular people at a particular time.

In the second part of his methodological doctrine, Savigny presented the important for the further development of German legal science differentiation between historical connexion and historical separation.

According to his idea of the development of the system as a whole, historical connexion is not limited to the research of the problem how a particular question has been answered through the time by particular legislations. Rather it takes care of the presentation of the system of law in its historical advance.

Historical separation gathers an importance in cases when two sources are present which claim to be both all-embracing, e.g. Roman and German law. However, historical separation means not the absolute and complete separation of jurisprudence from its historical sources, but rather hinders an undue mix up, i.e. when one "believes that the German [legislator] begins where the Roman one has stopped". This historical separation which at the end should supply the judge with a "practical result" requires, as Savigny himself recognized, the strongest efforts on behalf of the jurists: The contents of each particular legislation should be deduced and from that a theory of connection should be inferred. This idea determines also Savigny's notion of system, which represents not only an arrangement of legal sentences but rather their inner correlation. The latter is based on the mutual interrelations of single special laws on the one hand, and on the relation between rule and exception in each one of these subsystems, on the other. Thanks to this concrete and based on positive law notion of system, Savigny could dispense with all other pure systematic-philosophical models, and particularly with natural law. As Savigny himself suggested, the very function of the latter is a "kind of a revolution against legislation" however realized not as an open rebellion but rather "indirect; hidden" in the form of the severe axiomatic system: "Jurisprudence in itself could be studied with or without natural law".

Jurists do not need philosophy, and those who are not driven to it should better forget it (Savigny, 1951, p. 50).

The question whether Savigny himself remained true to his own methodology in the course of all eight volumes of his main work *System des heutigen römischen Rechts* ("System of Modern Roman Law") is disputable (See Wieacker, 1967, p. 397).

3.2.2.3. *Jurisprudence of Concepts*

Savigny cannot be considered a representative of jurisprudence of concepts. However with his fundamental idea of the legal system as an "organic unity" (Savigny, 1840, p. 75) he prepared the way of this particular methodological trend. This idea leads at the end to the conclusion that law as a system is complete in itself. Savigny required such a completeness of each code of law (1840, p. 21). This completeness could not be 'material' – encompassing all real legal cases, but it is rather a "completeness of another type" which Savigny explained with the help of a geometrical example. According to him in each triangle there are particular specifications the interrelations of which determine necessarily all others e.g. the triangle is defined by its two sides and the angle between them. In a similar way, each part of law contains stipulations, which determine the whole of it. Such stipulations are called leading principles. The based on these principles procedure of determination of all interconnections and types of relations between juristic terms and sentences, on the one hand and of their presentation with the means of the very system, on the other, is called by Savigny "organic complementation" (1840, p. 74).

Georg Friedrich Puchta (1798 – 1846) was Savigny's successor both on his chair and in his leading position in the historical school. Puchta took up Savigny's main idea about the "organic correlations between all legal sentences" which could be explained historically as resulted from the *Geist des Volkes* (spirit of the people). However Puchta attributed these correlations to logical reasoning and not to the *Geist des Volkes*: The sentences of law are organically interrelated in a whole on the basis of the very nature of law and of the rationality law is confronted with. What is more the connexion between the sentences acquires a systematic character: They pre-condition and presuppose each other – the existence of one sentence makes possible to conclude about the existence of the others

(Puchta, 1893, p. 21). In this spirit Puchta defined methodologically the subject-matter of legal science, namely to determine the systematic connexion between the legal sentences, to trace the genealogy of each one back to its principles, and *vice versa* – to deduce even the last sentence of the system from its first principles (Puchta, 1893, p. 22).

Savigny's "organic complementation" is transformed by Puchta to a "product of deduction". According to him only a scientific deduction could reveal those legal sentences which hidden in the "spirit of the national law" are not to be met either in the immediate convictions of the people or in the legislation (Puchta, 1893, p. 22). Thus Puchta turned legal science into a legal source itself with the *Volksgeist* and legislation together. The specificity of this third legal source is that only it – thanks to its systematic character – could guarantee to encompass all parts of law. Puchta illustrated the particular fundamental function of science to fill systematically gaps by comparing it with a piece of art: If we consider law as a mere aggregate of legal sentences we would be never able to recognize that a part is missing whether we could always determine a missing part in a piece of art. Puchta, 1893, p. 57). This comparison of the legal system with a construction refers only to the outer, formal structure and not yet to the inner correlations. It is however only those who possess this systematic knowledge about the inner correlations between the legal sentences could trace "upwards" and "downwards" the genealogy of each term on the basis of all "intermediate members" which participate in its formation. (Puchta, 1893, p. 57) Puchtaclarified his statement on the example of the passage servitude – the right to pass an estate given by the owner to neighbours. In order to stipulate this particular act the jurist has to define its position in the whole system of legal relations namely to trace this particular act back to the term of law in general, and from there – to the particular act again. The passage servitude could be analysed as follows:

- It is a right to a thing,
- A right to a thing which belongs to somebody else, e.g. only a partial subjection of the thing.
- This partial subjection refers to usage, e.g. a right to use things.
- The usage is ruled for a particular subject, e.g. it is a servitude.
- Which refers to an estate, e.g. praedial servitude.
- For this particular need concerning the estate, e.g. a passage servitude.

Puchta called this structure a "genealogy of terms" and underlined that it should not be perceived only as a "mere scheme of definitions". He saw each term as a "living being", as an "individuality" different from the individuality of its progenitor. Puchta's genealogy of terms is not a formal-logical pyramid of terms but rather a developmental theory of terms taking account of form and content: "The father of legal institutes[5] (Puchta, 1893, p. 8) is the principle of law, the mother – the subject-matter, i.e. the manifoldness of people and things" (Puchta, 1893. p. 57).

With his genealogy of terms, Puchta advocated not so much a formal-logical truth-claim but rather a productive developmental theory of law. His jurisprudence of concepts is indeed only a construction-method subjugated to the "inner logical approach" but no way a formalism in the sense of modern logics. That's why Puchta's deduction of terms possesses a heuristic and creative character bringing to light legal sentences omitted by the *Volksgeist* and law altogether. This procedure, later to be censured by jurisprudence of interests as an "inversion-method", accounts for the contents of the basic principles and is of no formal character. Puchta inferred the development of law "from the term" which is more than a consequent turn towards the systematisation of juristic dogmatic aimed at the transformation of positive law into a demonstrative science. The constructive method characterised the whole pandects-science of the 19th century. Thus it was further developed by Ihering, Gerber, Laband to reach its completion in the work of Bernhard Windscheid who penned the most meaningful textbook on the pandects. The lack of codification of the valid law made the book a scientific authority in practicing common law (Wolf, 1951, pp. 584-586). As a member of the First Commission on civil code Windscheid marked the end not only of the pandects science but also of the classical juristic method. In the pandects science the classical method had as a central problem the filling of the existent gaps, its main goal being the completeness of the system. According to the theories of Puchta and Windscheid the logical inter-connexion of terms in the system permitted the consistent filling of gaps in positive law by means of the so-called "productive construction". The latter served the "filing out and the polishing of legal scientific terms till complete justification of the system" (Wieacker, 1967, p. 436), practiced unrelentingly by the pandects scientists. Hence the "juristic method" helped legal scholars and not practicing jurists. The aim of the former was to complete the exploration of legal terms and thus to make possible the logical subsumption of whatever legal case under a proposition or term. However, this necessarily leads to the limitation of the judge's ruling to the logical act of subsumption. This ideal picture of the juristic method was adequate to the legal science of the period pursuing to fit the still valid Roman legal sources to a logically indisputable system of terms – the so – called pandects law. However radical changes in the jurists' methodological orientation had been

[5] = the sum total of legal sentences of a particular law, e.g. property (Puchta 1893, p. 8).

observed already before the introduction of the Civil Code on the 1st of January 1900. They were provoked not so much by the gathered popularity speech of the prosecutor v. Kirchmann directed against the self-complacency of juristic theorists but rather by an internal transformation of the pandects science itself. Kirchmann, J.v. indeed recognized the fatal cleft between speculative legal science, on the one hand and the political administration of justice on the other. He exposed the precarious situation of pandects science in the modern constitutional state which situation remained concealed as long as the governments and parliaments let the codification in the hands of scholars. (Wieacker, 1967, p. 415) Kirchmann however was not able to offer a way out of the situation. This was done by Windscheid's friend and contemporary Rudolf von Ihering, who felt close to the spirit of Roman law. Ihering's activity was marked by two phases in the course of which he completed the transition from jurisprudence of terms to jurisprudence of interests, which brought him the name of father of the newer methodology (Kriele, 1967, p. 63).

3.2.2.4. *Jurisprudence of Interests*

According to the historians of law, in the first phase Ihering (see Wolf, 1951, p. 616) pursued further Puchta's jurisprudence of concepts in the extreme style of the natural-historical method. This method had been already criticized, and in his second phase Ihering himself describes it as a "catchphrase elevated to a scientific method". What has been mostly attacked – actually till nowadays, is the apparently indiscriminating application of pictures, which exclude each other, e.g. "organism" and "machine", or the application of pictures from various spheres of natural sciences. As an example one could use the comparison of law with a physical body

- Possessing the property of impenetrability. Thus a law cannot be at the same time both in A and B not even in the course of a "juristic second" but either in A or B: A negated law could thus not be once again negated.
- Or, possessing the characteristics of elasticity according to which a property goes back to its old size after cessation of the burdens (Quoted after Wieacker, 1967, p. 434).
- Or, possessing chemical properties as a result of which legal science turns into a "juristic analytical chemistry seeking simple bodies". With the help of juristic construction the given "raw material" (the legal rules) is "dissolved to terms" which act mediates the transition of law "from a lower to a higher state of aggregation" (Quoted after Larenz, 1979, pp. 28-30).

Despite all criticism against those picturesque analogies, one should consider (as Wieacker himself correctly stated) the heuristic and creative functions of such an illustrative approach to factual and procedural fundamental relations in legal methodology. And indeed Ihering realized a general model of juristic cognition and decision-making on the basis of analogies between relations of the real-world things and events on the one hand, and relations of the stipulations of the legal "ought", on the other. This model is amply presented in the second volume of Ihering's main work *Geist des römischen Rechts auf den verschiedenen Stufen seiner Entwicklung* (The Spirit of the Roman Law at the Different Stages of Its Development). The model of "juristic technique", as Ihering called it, consists of three fundamental operations:

- Juristic analysis
- Logical concentration
- Juristic construction

According to Ihering the juristic analysis represents in itself the decomposition of the existent (positive) legal material into its basic elements. The result is something like a "legal alphabet" or a limited number of juristic basic terms. Their combination construes the variety of concrete legal relations in the real world.

This intellectual operation, namely the abstraction of the general from concrete cases embodies already one of the applications of the analytical method. The latter is nothing else but a "separation of the general from the singular, the decomposition of the material into its general and particular or local components". As examples for such general questions concerning all types of contracts Ihering offered, e.g. the problem of the impact of error on the validity of the contract, and the problem about the consequences of diversity or of delay of performance (Ihering, 1852-63, p. 336).

Ihering viewed abstraction as the most important application of analytical method. He claimed that jurisprudence as a legal science could only "extricate" the general and not create it. According to Ihering, this is a fundamental rule of human knowledge: In each sphere of knowledge the human mind perceives first the concrete and later the general. The concrete parts in law i.e. the legal sentences reflecting single legal relations appear to be earlier developed historically as the abstract parts. Not rarely the latter have to undergo different developmental phases and a long pre-history before legislation or science could recognize and formulate them in their true i.e. general form (Ihering, 1852-63, p. 338). The single phases in the development of law take place as successive, step by step abstractions. The common among the single legal relations reveals itself in a particular, historically precisely defined place. Ihering called this place the "historical break-through" place. In Roman law, for example, ship-business and trade (*act. Exercitoria* and *institoria*) proved to be the historical break-through place for generalization of the idea of principals' discharge from liability for agents' contracts.

The historical break-through ensures the initial generalization of single, specific legal relations. However, at this stage the general legal sentence has not yet reached its "true shape". This last generalization absolving the legal sentence from its specific and historically determined forms of appearance is achieved through the so-called operation of analogical extension. According to Ihering, the operation of analogical extension represents no interference on behalf of jurisprudence as a legal science in the rights of the legislator. Rather, it is to be seen as a higher critique and interpretation of legislative thought.

The "analogical extension" is no arbitrary operation. Rather it is subjected to the general developmental law of human cognition, consisting in a "slow maturation of ideas". The necessity of such an analogical extension has to be universally recognized – the efforts of "single individuals will have no consequences" (Ihering, 1852-63, p. 343). The analogical extension is substantiated by the very "nature of the thing" i.e. it takes normally place when real relations necessitate the generalization of particular cases provoked by new cases. For example, the old law on counterfeit money refers to coins. With the appearance of bank-notes the old law had to be extended to paper-money which was argued in the following way: The law viewed the generic notion of money (the publicly attested means of exchange). This law however was attached to the only art of money existing at the time of its introduction – metal-money. The essential in it however accounted for the class (money) and not for the art (coins) (Ihering, 1852-63, p. 33).

Ihering claimed that juristic analysis, representing a at least two-level abstraction process was the scientific method of jurisprudence. Its application could show the difference between an expert and a layman. A layman's idea of juristic analysis is antipodal to the one of an expert. The former is guided by an "instinctive all-out feeling" which could lead to a disadvantageous choice of lawsuit. In such a case the judge would be forced to reject it although with a different lawsuit the plaintiff's claim would have been valid. Therefore Ihering viewed the "organic" idea of law (represented in German legal philosophy mainly by Stahl) as lay and non-scientific. He actually quoted a friend of his (a jurist) who used to say mockingly: "What one cannot define is described as organic" (Ihering, 1852-63, p. 352).

The scientific method opposed to analysis is not, according to Ihering, the organic approach but the logical concentration. However the latter is not a specifically juristic operation but the logical operation of "abstraction of a principle from facts". In that case Ihering didn't refer to the formal-logical class abstraction (notion's extension-content relation). Rather he attempted to set out the essential in the sense of an "intensification of the logical terminology". That could be seen in Ihering's conception of a model, which resembles very much the contemporary theoretical idea of a scientific system. Ihering claimed that legal sentences reflecting an unconscious application of a principle on behalf of the legislator related to this principle as the single points of a circle to its centre. Before the legislator could find the principle he looks for, he is forced to describe it with the help of specifications of the particular case. In search for principles science follows the same pattern – the further the distance to the centre is the longer is the way science must go and the more complicated its explanations of a particular object are. Each step nearer to the centre decreases the number of its propositions and enriches the content of the latter. In reaching the centre science is able to encompass the whole matter in a single thought (See Ihering, 1852-63, p. 354).

As a general method logical concentration prepares the transition to the very juristic operation – the construction – bringing the analysed material to a systematic unity. As Ihering correctly stated, already during his life-time juristic construction had been a currently used conception in jurisprudence. However it lacked a strict definition not to mention a theory. Ihering tried to fill this gap by a better definition and by the formulation of the laws of juristic construction. According to him juristic construction represents the organisation of the legal material into a juristic body. Ihering used the notion of juristic or legal body in the sense of Puchta, G.F.'s legal institute i.e. the sum-total of legal sentences of a particular law e.g., property law or mortgage law.

The technically correct organisation of the legal body follows particular laws of juristic construction:

- Juristic construction has to cover all positive material: The positive legal sentences should be all encompassed by it.

Otherwise juristic construction is completely independent – also in regard with the legislators' constructions which are only of doctrinaire importance and could be anytime revised or disposed of by legal scientists.

The organisation of those legal relations could be realised only through different juristic constructions: The construction that property could be ruined and then again revived, and the construction that property indeed persists but could not be set into operation as long as the connection exists. Both constructions are possible according to the above-mentioned law. The first one however is imperfect compared to the second. And although the legislator himself could have used the first construction in the particular law it could be exchanged for the second one by the jurists. The result would be practically the same.

In this spirit one has to interpret Ihering's general definition of the notion of juristic construction: Juristic constructions represent the attempts of science to clarify rationally sentences accounting for the legal practice. Ihering shared the opinion that the legislator should not participate in the process of construction. By doing so the legislator would only encroach upon the sphere of science and would damage his own legislative authority and power stepping down to the position of a jurist.

- The law of non-contradiction or systematic unity: This law refers not to inconsistencies on behalf of the legislator but rather to contradictions in the legal science itself i.e. the constructions of science should be in "tune" with themselves and among themselves.

Accordingly Ihering formulated the corresponding law as follows: Science should not constitute juristic impossibilities.

The notions of juristic possibility and impossibility are for Ihering not absolute but relative conceptions, which undergo changes in the course of legal development. Both law and science are marked by a constant progress widening their spiritual horizon and therewith the range of possibilities due to their own efforts or to the fact that what has been considered as juristically impossible realises as practically necessary and thus stimulates the extension of the sphere of theoretical possibilities (Ihering, 1852-63, p. 377).

On the basis of this historical relativity Ihering argued the existence of more or less perfect juristic constructions, which bring the logical and the positive elements together. The third law, being however not on a par with the previous two ones, is called by Ihering

- The law of juristic beauty, which refers to the level of perfection of juristic construction.

The criteria of perfection is the construction simplicity – a concept used from antiquity till nowadays for the evaluation of scientific theories without being ever strictly explicated.

The presentation of the laws of juristic construction marks the last stage in Ihering's concept of juristic method. Ihering himself recognized the incompleteness of his model: Thus the organised as a system – with the help of juristic construction, – law is not only the most advantageous form of positive law but at the same time an "inexhaustible source of new material". Therewith Ihering entered the second phase in his work – the intent-methodologically oriented one indicating the consequent transition from jurisprudence of concepts into a jurisprudence of interests. The juristic construction presents indeed law as a system. However the moving force behind it stimulating which stimulates the progress of law as a system is the "philosophy of practical intent". The latter is able to explain the development of law not only formally but also with respect to its contents.

Ihering's second phase in his activity dedicated to the philosophy of practical intent is programmatically formulated in the first volume of his work *Der Zweck im Recht* (Law as a Means to an End) published in 1877. According to Ihering the maker of the whole law pursues the following goal: There should be no legal sentence, which doesn't owe its existence to a particular purpose, to a practical motive. (Ihering, 1893, VIII) In his study Ihering explored the question about the intent-subject who as a real person set up the legal sentences and provided them with validity. Ihering could not be any longer satisfied with the simple answer defining that subject as legislator (see Larenz, 1979, p. 50) for he had already recognized the social character of the concept of purpose – the latter could be generalised only if it expressed the common purpose of all members of a community. In this sense the legislator represents not a single person but rather the society. Hence all legal sentences serve the purpose to guarantee the living conditions of the society.

With those considerations (which he methodologically didn't evaluate) Ihering prepared the start of the methodology of the so-called "jurisprudence of interests" developed further primarily by Philipp Heck. Heck coined the term "concept jurisprudence" to name the methodology used by Puchta, Windscheid and by the early Ihering (See Larenz, 1979, p. 53). In the juristic methodological doctrine he is known for the formation and evaluation of the conception of closed system (See Esser, 1956, p. 222). In Heck's opinion the central problem of the methodological dispute concerns the "influence of law upon life and the way case rulings mediate this influence" (Heck, 1932. p. 4). He maintained that the methodology of jurisprudence of interests related only to the "practical legal science". In other words, it "facilitates the functions of the judge and it prepares the factually adequate decision through the exploration of law and life circumstances" (Heck, 1932, p. 4). The concept jurisprudence is concerned only with the "outer" order of positive law thus possessing merely a descriptive value. The jurisprudence of interests pursues the knowledge of the "material, national, religious and ethical" interests behind each legal concept and legal sentence. Law in its essence is nothing else but "protection of interests" and the legislator – a kind of "transformer" (see Larenz, 1979, p. 55) or "collective symbol" of the interests of the legal community. These legal-theoretical considerations promote the most fundamental methodological claim to "recognise historically correctly the real interests behind law and to take account of them in the process of case-ruling" (Heck, 1932, p. 60).

This fundamental claim of jurisprudence of interests serves to explicate a crucial differentiation in juristic methodology namely the differentiation between the logical-systematic argumentation of law, i.e. the subsumption of a particular case under the legal sentence (or in terms of the concept jurisprudence – under a hierarchy of legal concepts), on the one hand, and the finding of law in the sense of an adequate to the interests assessment of the case, on the other. Heck as well as the later representatives of the jurisprudence of interests didn't totally reject the logical-systematic argumentation of law. Rather they perceived it as a reversal of the factual process of finding of law, which could lead to the correct decision if the legal conception had been formed on the basis of a precise assessment of the interests. In such a case the logical subsumption would be not only a useful but also an indispensable instrument for "steadiness and promptness of the administration of justice". The assessment of interests has however an absolute priority over the formal-logical operations. Thus each concept is nothing else but a symbolic expression of scientific consensus about the general assessment of interests and conflict decisions of the legislator, the latter being in his turn only a symbol of community interests. This approach reveals a fundamental weakness of the older jurisprudence of interests:

The concept of interest itself has a triple meaning – of an object of assessment, an assessment standard, and a causal factor. Its function as an assessment standard is determined by the fact that the general ideas of certainty of law, of justness etc. are perceived as "ideal interests" themselves to be evaluated by the legislator as well. In the spirit of the genetic jurisprudence of interests Heck maintained that law was "caused" by real, historically dependent interests of the society – hence the function of the concept of interest as a causal factor. The so-called "jurisprudence of valuation" tried later to compensate that lack of differentiation of the concept of interest. It distinguished between interests in the true sense as desirable goals of the parties in a litigation when the former aimed at beneficial for those parties legal consequences, on the one hand and assessment standards which were not interests themselves but rather inferences from the idea of justice, on the other (Westermann, 1955, pp. 14-16).

The change from concept jurisprudence to jurisprudence of interests marks the end of the so-called "traditional methodological doctrine". It shows that the development of juristic methodology takes place in a succession of stages, which correspond to the developmental stages of the legal system representing the sequence of stages in the development of human awareness of freedom. The stages in the methodological development are to be found – although not always complete –, in all historically determined legal systems. For the development of law as well as the development of juristic methodology has so to say spatially wandered from one people to another.

- The origin and the first stage of juristic methodology is to be found in the ancient civilisations and in the ancient Roman jurisdiction. Typical for this stage was the inductive case law or casuistry i.e. ruling from case to case. Analogy played an important role in case law supplemented occasionally by the formal method of interpretation, which softened the severity of ancient law (e.g. the twelve-tablets-laws of the Romans).
- The second stage was characterised by the introduction of the hypothetical concept construction, which had already been used occasionally by institutionalised authorities (jurists, judges). Crucial for that stage had proved to be the personal knowledge and logical-argumentation abilities of those institutionalised authorities. The origination of legal science is traced back to this stage.
- The last developmental stage of the legal system – the axiomatic-deductive one (which could be best compared to the German jurisprudence of concepts) has remained an utopian ideal. For it has to represent a gapless, logically closed legal system in which all recognized laws could be derived from a few axioms.

To sum up one has to say that the growing complexity, diversity and organisational level of the legal system should correspond to a increasing efficiency of the methodological technique in mastering legal circumstances of ever increasing complexity. The latter is determined by the growing awareness of man of his own freedom. This supports the idea that law is nothing else but the "deed" of humans i.e. made wholly by individuals who as a result have to carry the whole responsibility as well (Wadle, 1982, p. 10).

3.3. *Extension Towards a Legal-Ethical Theory of Principles*

Juristic theory of argumentation makes a distinction between

- Internal justification, which deals with the question, whether or not the judgement logically follows from the premises of its foundation, and
- External justification that refers to the foundation and completeness of the premises themselves.

External justification is related in various ways to the theory of transformation, since it basically includes rules and forms of justification, which – unlike internal justification – cannot be presented in a deductive-logical manner. While internal justification is sufficiently grounded in the general rules and techniques of conclusion, external justification – according to Alexy the "very field of juristic argumentation" – requires particular rules and forms. These forms are (See Alexy, 1983, pp. 273-275) rules and forms of:

– Interpretation,
– Dogmatic argumentation,
– Utilizing prejudication,
– General practical argumentation,
– Empirical argumentation,
– So-called special juristic types of argumentation.

The transformation-theoretical pendant of these rules is the systematisation of legal sources as "must", "ought", and "can" However, the first of all transformation is the "leap" into law; it is the pre-condition of all, even the most simple internal justification which treats universal premise as valid law.

The leap into law, i.e. the incorporation of extralegal principles, is substantial to the juristic practice and not just a problem of legal philosophy, although practical jurists are rarely concerned by it. Theory of transformation in the sense of Pecznik has one fundamental disadvantage, which, however, might be corrected. It brings the context of discovery entirely to the empirical sphere of psychology, which methodologically remains unanalysed. From the point of view of today's philosophy of science such a renunciation of explanation is unjustified, the more so, since contemporary philosophy of science offers possibilities of an extending rationalization of the so-called context of discovery *sensu* logics of discovery. Habermas' concept of evolutionary "logic of development" is to be understood as a hint at such a logics, which essentially consists of the reconstruction of the developmental stages of law that determine the basis of the leap into law.

This basis is the respective moral order. It was not given at the beginning, but has developed from the low level of moral sense to the legal-ethical principles. Juristic theory of argumentation, if it should not be just a juristic technique, but is expected to meet generally and fundamentally the standards of the methodological dimension of legal philosophy must be extended. Its extension refers to the legal-ethical principles. Legal positivism attempted the elimination of these principles and separated law from morality. The inappropriateness of this attempt was known already at the time, when legal theory emerged. Thus, in 1926 Burckhardt – an advocate of the Neo-Kantian approach to legal science –expressed his doubts in the then newly founded *Internationale Zeitschrift für Rechtstheorie*. The founders of this journal were Kelsen, Duguit, and Weyr. Burckhardt expressed his doubts against a project that excluded all problems of "justice, right, just, natural or absolute law" and accepted only a theory of positive law[6]:

> The gentlemen cannot simply as they like define and state: this is meta-juristic, I do not want to get involved in this. In science there are interrelations between everything, interrelations between law and legal philosophy, legal philosophy and general philosophy. Whether or not one likes it, only the person with the broadest overview will obtain the well-established answer; contrary to arrogant scholarship and sciolism, science has a difficult task (Quoted after Dreier, 1981).

This criticism was admitted, to a certain extent, by Kelsen, who later conceded that his *Reine Rechtslehre* is not that "pure" (as a mere epistemology of law) (see Raisch, 1988, p. 15), because his "basic norm" – presupposed by a positive legal order as a reason for validity – has some similarity to natural-law theories. But in addition, he stated: "This, however, is the only point", because "the basic norm determines only the reason for validity, and not the context of validity of positive law." Therefore, the basic norm is no "norm of justice" (Kelsen, 1960, pp. 442-443).

In the juristic theory of argumentation, this retreat to mere validity is not possible, for this theory adheres to the postulate of saturation of arguments. This saturation means that the foundation of any juristic decision has to include all arguments – indeed in an explicit and "transparent" manner – arguments that, according to the positive legal system, are relevant to a "just" decision.

Thus, this argumentation-theoretical conception of justice renounces, after all, non-transparent traditional phrases, which (in most cases) are unnecessarily superimposed to the very juristic foundation (Raisch, 1988,

---

[6] Burckhardt's letter to A. Gysin from 31.07. 1926

pp. 39-40). Such a traditional phrase is the old and venerable and often discussed "nature of things" (Raisch, 1988, pp. 44-45).

Using this phrase without any further specification and concretisation is indeed harmless but trivial, and therefore of no relevance with respect to the theory of argumentation. The same is true to some of its metaphysical versions, especially when they represent mere dogmas or ideologies. A negative equivalent of this traditional concept of "nature of things" in the modern theory of argumentation seems to be the notion "form of life". Basically, it indicates only in a negative sense the natural constraints of the practicable legal regulation. Yet there exists also a positive and rationally explicable interpretation of "nature of things" from a descriptive-empirical or prescriptive-normative viewpoint. However, (as Engisch himself explicitly stated), in both cases, a positive interpretation of this concept includes the possibility that "evaluations, setting of purposes, interests, etc. as facts pre-determine the correct law" (Engisch, 1977).

Such a priori given rules do not presuppose a legislator because they simply follow from the natural and historical development of humankind. They can be found in all spheres of human desires, needs, and purposes: Survival, nutrition, housing, clothing, reproduction, communication, information, free opinion, matrimony, family, obligation to pay alimentary, social security, etc.

All these examples make obvious that the legal regulation corresponds to pre-given structures of order. This does not mean that these structures – even if they are indeed "natural" and "adequate to facts of life" – cannot be transcended by independently established norms. Irrespective of the "super-norm", i.e. "a person should not be obliged to do what he/she is unable to do" (Moritz, quoted after Engisch, 1977), the very progress in the development of legal systems has always been conducted by the imagination of an ideal "pre-existing system of norms", which, however, requires a profound explanation – the hint at empirical and historical facts is not enough.

Now we have a basis for the connection between an argumentation-theoretical concept of justice and the classical idea of "just law" or "law of reason".

3.3.1. *Rehabilitation of Law of Reason*

If we stick to the metaphor "leap into law", which is suggested by transformation theory, then we have first to ask which of the possibilities offered by traditional legal philosophy can be used as a basis for the "take off". From the point of view of argumentation theory and its conception of justice, this question has been answered by Dreier in an unambiguous manner: "From law of reason to positivism and back again".

At first glance, however, the rehabilitation of the law of reason is grounded in negative aspects: In spite of enormous efforts, legal positivism could not replace the traditional law of reason, which is mainly based on Kant's philosophy. Instead of yielding more legal security, the basis of foundation theory deteriorated in a catastrophic way (See Bydlinski, 1988). On the other side, a rigorous theoretical a priori – including content-determined legal-ethical principles with the features of absolute validity and stability, and tied to the claim of complete recognition – is no longer tenable. Kant did not advocate such a view, which made itself clear in the formalism of his ethics. What could be absolutely detected is merely the formal structure of a generally valid legislation. The latter is expressed in the categorical imperative from which the supreme legal principle can be deducted.

To overcome this often criticized formalism – Kant himself attempted that in his works on philosophy of history – one has to go back to the origins of natural law in Aristotle. Apart from some pure naturalistic attempts in the Sophists' philosophy, classical natural law has always been law of reason. As a generally valid, necessary knowledge about the essence of justice, it has also been a priori, i.e. demonstrable independently of experience:

- In the sense of Christian natural law (Th. Aquinas) it is deductible from the revelation of Divine law.
- In the sense of natural law in modern times (Hobbes) it can be gained from the human free will that itself creates the principles of justice and injustice.

Moreover, Aristotle presented a still up-to-date solution in the context of discussions concerning "law and practical reason". His approach can change legal philosophy into a theory of practical reason. In his lectures on ethics he stated that there was also a practical truth (1139a, pp. 26-27).

The meaning of this seemingly trivial insight becomes obvious, if one compares jurisprudence as a practical science with medicine. According to Aristotle (1105b, pp. 12-18) most people take resort to theoretical speculations believing that in this way they become competent. They are like sick people who think that in order to recuperate it is sufficient to pay attention to what a physician has to tell them.

Theoretical reason, then, does not suffice when it comes to practical problems. Conversely, practical reason does not renounce theory and each relation to truth. Rather it has its own way to truth independent of theoretical reason. This way, however, cannot be easily found. Aristotle himself maintained that only the confirmative statement (*logos apophantikos*) **can** be true or false. It can be true either by stating empirically testable facts or by allowing for prognoses, which later can be verified or falsified.

Yet it is not at all clear, in what aspect non-empirical and non-theoretical practical sentences like a question, a request, a prayer, a demand, a command or a prohibition – and therefore law in general – can be related to truth. (Inciarte, 1974, p. 158) Aristotle said that such sentences in fact do not lead to theoretically recognizable truth, but that they make sense, because they are to be realized only by concrete action. Practical reason does not only mean to recognize truth from an external point of view, but to cause truth from inside, as it were. It is connected to the idea of an operative and not merely contemplative truth. Operative truth finds its expression in good and just action, in order and law. Since a merely theoretical reason deals only with the existing and is "looking" at it passively, it is not apt to cause something good. "For itself – i.e. theoretically – reason does not move anything" (Quoted after Inciarte, 1974, p. 162).

The Aristotelian doctrine of the internal operative "truth capability" of reason corresponds to the more recent Kantian doctrine of "inner legislation". As Dreier (1986) made evident, Kant's concept of law comprises "natural law" – Kant himself used this term though he should better have spoken of "law of reason" – as well as positive law. As he bluntly remarked, a "merely empirical theory of law" would be a head without a brain (Quoted after Dreier, 1986, p. 9).

Hence, practical reason – and not just some authority as legislator – also determines positive law. According to our present day's view, a judge's decision should therefore be understandable by means of practical reason. A person should recognize that not arbitrary use of power, but rational argumentation determines the legal decision (See Raisch, 1988, p. 22). In the sense of Kant this means that a judge's decision refers to "practical reason": Everything in nature is bound to laws. Only a being endowed with reason is able to act according to the imagination of laws, i. e. to act according to principles, or is capable of will. Since the deduction of actions from laws requires reason, the will is nothing else but practical reason (See Raisch 1988, p. 22). The validity of decisions is based on reason, and the decision itself is to be regarded as "turn of the obligation to explain" (Kriele, 1979, p. 40). However, in today's perspective (Kriele, 1979, p. 41) one cannot suppose that all arguments and counter-arguments can be completely discussed with ever-lasting results. This does not mean that the rationality of the decision is called in question. In many cases it is reasonable to decide, after all, and not to renounce any decision. It is enough to decide according to the most profound arguments. But such decisions can always be given up again.

Positivist legal theory negates both good arguments for decision and profound arguments for its revocation. In the spirit of the classical imperative law theory – which regards laws as human commands – it has always connected decision with the notion of pure subjectivity, arbitrary act or "blind decision". In the sense of an "enlightened" (Kriele) legal philosophy as theory of practical reason, decisions express results of arguing and must therefore be well-grounded. This is not true only to positive law in general, but also to single legal norms, which are obligatory only if they themselves can be accepted as ethically founded (Kriele, 1979, p. 120).

The legal-ethical principles, on which each just decision is based need not and cannot be always made explicit. As operative truth practical reason is more rational than it could be when presented theoretically or even from the point of view of language analysis. In this context, the analytical discourse only supports and stabilizes the practical one. This means: In the case of logical/terminological confusion it must be always possible to move from the practical discourse – which refers to the reasons for a decision with regard to contents – to the analytical discourse and immediately back again. The analytical discourse does not stand for itself, and after all the whole methodological dimension of legal philosophy represents only the *condition sine qua non* of the practical process in search for justice. This is also true to the whole corpus of deontic logic, according to which reason can indeed be practical, because it is not senseless to norms. Nevertheless, Wright (1963, p. 104) proposed to keep the question of truth in distance to norms, because they do not represent facts. According to deontic logic the question of truth refers only to meta-statements, i.e. statements concerning norms. These statements themselves are no norms or practical rules.

However, an argumentation-theoretical conception of justice relies to those fundamental principles, which are practical statements of legal ethics and are implied in each just judgement. Deontic logic might be important, but it is not sufficient to meet the practical requirement of truth, which has to be a characteristic of legal-ethical principles. A norm or law by itself is neither true nor false. True or false are our statements, while a law simply exists (or does not exist). Hence, in the context of deontic logic truth in relation to law does not mean the practical truth content of the law itself, but merely the content of theoretical statements about law, especially the question of the "ontological status" of a norm, i.e. in the case of law the question, whether and when a particular statute exists (Wright, 1963, see also Inciarte 1974, 165).

What, now, is a law's practical value regarding truth? The traditional answer is: It is given by its correctness (*rectitudo*) or incorrectness.

This viewpoint is still the basis for the axiological dimension of legal philosophy as the fundamental quest for correct or just law. The answer to this question has to be regarded (in the sense of argumentation theory) as the most thorough, substantial foundation of a juristic decision. It is to be seen under the premise of practical reason, and its solution is still offered by the traditional applicative circle of practical jurisprudence which can be represented by the model of a judge: "Legal science should not operate in a manner different from the judge's operation" (Raisch, 1988, p. 78).

3.3.2. *The Model of Deeper Juristic Argumentation as a Special Case of the Applicative Circle*

The starting point is the classical distinction between theoretical and practical sciences. The former are based on theoretical, the latter on practical reason.

Theoretical sciences attempt at recognizing reality, practical sciences try to master and to change it. Therefore their methods are also different. In the descriptive-theoretical sciences, the so-called hermeneutical sciences (e. g. philology and historiography) and the scientistic disciplines like natural sciences, the methods are methods of knowledge, while methods in the practical sciences (to which traditionally medicine and jurisprudence belong) are methods of application. Within comparative typology of the sciences, the argumentation-theoretical model of deeper foundation is nothing but a special case of the so-called applicative circle, which differs from the mere interpretative hermeneutical circle because of its prescriptive or normative character (See Gadamer, 1965).

The basic model of the applicative circle is rooted in the hermeneutical tradition. However, unlike the model of interpreting texts, it has always been oriented towards the model of the judge or the physician. It can make explicit that the justification of a judicial decision has a "fairly more important function than a justification based on philosophy of science" (Raisch, 1988, p. 20). Philosophy of science is a justification *a posteriori* and represents a rational reconstruction of past knowledge processes. In jurisprudence justification has to precede each decision. It does not follow the decision, but the decision is based on it, no matter that in the practice in the court it is formulated after the judgement. The obligation of justification is linked with the compulsive decision, which includes also a component of time. This can be shown even clearer by the model of a clinical decision, which gives a structural and functional equivalent to judicial decision. In the case of an acute illness the physician is forced within a certain period of time to decide for a particular therapy, for otherwise the patient will die and the case will solve itself.

A legal case is similar: One has to find decisions, even if one has only incomplete information and has not discussed all pros and cons in detail. "Non-decision" always means injustice. The right decision in a legal case is therefore always a case for practical reason.

The structural equivalence of both models of decision, the model of the physician (Fig. 1) (see Schwanitz, Oeser, 1983, pp. 291-293, Oeser, 1988) and the model of the judge, can be illustrated as a cycle of foundation (diagnosis, norm hypothesis) and decision within an explanatory (nosology) or a normative system (code) as special cases of the general applicative circle.

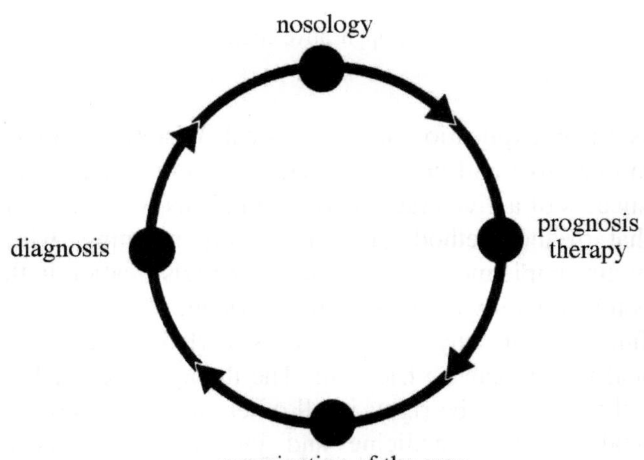

Fig. 1.: The applicative circle in medicine

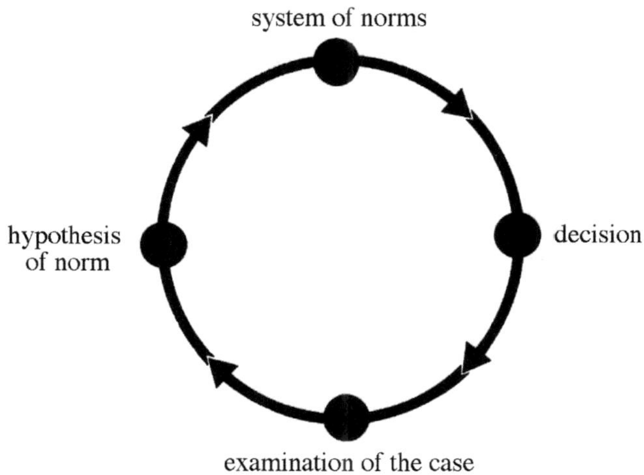

Fig. 2.: The applicative circle in jurisprudence

Both cases are an explication of the classical Aristotelian notion of practical truth as an operative and procedural view of truth that can be tested by the practical success of a given rational decision. This way, the notion of truth as well as that of the method gains a weighty meaning, which cannot be grasped by the simple model of verification or falsification in those scientific disciplines that aim only at theoretical knowledge.

The function of any theoretical knowledge, e.g. a physical or astronomical theory, can be tried out: The theory is verified if it allows for prognoses that prove to be right; in all other cases the theory is falsified. In the practical sciences, medicine and jurisprudence, this trial-and-error method is unwarranted, because a physician's or a judge's error means a wrong decision which in most cases cannot be simply corrected: for the involved persons, patients or clients (defendants) it can have irreversible consequences.

Hence, in practical science foundation and justification are more important than examination; they should even make the *a posteriori* examination unnecessary. Yet in medicine there is animal experimentation (a controversial issue, indeed), in jurisprudence there exist no "tentative decisions". In the best case, decisions can be theoretically discussed and their possible consequences critically reflected. The *a-posteriori* examination, the principal method in the theoretically founded empirical sciences, is replaced in jurisprudence by a hierarchical system of justification.

3.3.3. *The System of Argumentation Stages of Juristic Decisions*

Such a system has existed ever since jurisprudence was established as practical knowledge of decisions concerning law and justice. The classical canons of interpretation (see Raisch, 1988, p. 70) in traditional methodology have never been exceeded. They are still present in modern methodological approaches, like the juristic theory of discourse or argumentation.

If one now looks at juristic theory of argumentation in the narrower and wider sense as a special case of the applicative circle, one can incorporate in such a model also the procedures listed by Alexy as necessary for legal order:

- The procedure of creating law by the state (pr) is located in the first quadrant, i.e. updating law;
- The procedure of juristic argumentation in a narrower sense (pj) can be found in the second quadrant as primarily legal-scientific constructive task (legal dogmatics);
- The procedure of judicial operation finds its place in the third quadrant, where the judge makes his/her decision and subsumes his/her case under the constructed "system of axioms" of a code according to the rules of topic and founding law;
- In addition, the procedure of social efficacy (pw), which is independent of a judge's decision, shows the practicability of any applied legal rule (and can be verified or falsified, showing whether or not a given case can be subsumed under the law).

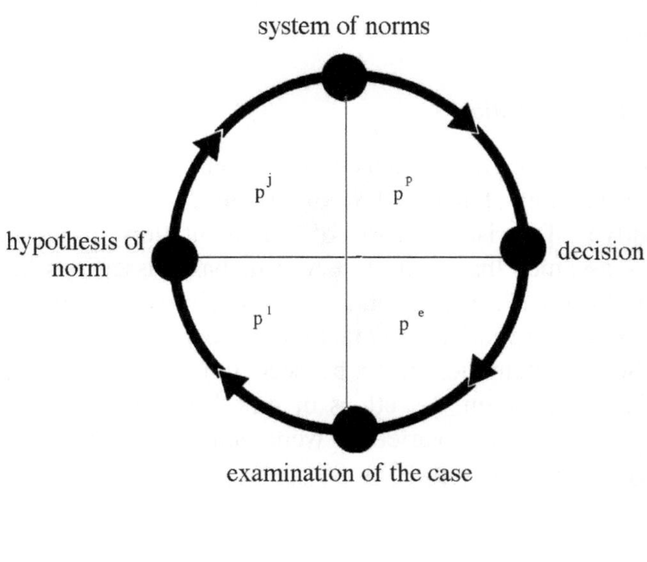

The further development of this model should make clear that the legal procedures logically presuppose the two procedures of creating law by the state and of juristic argumentation in a narrower sense – in a way, it has even to recapitulate them in an abbreviated form.

Therefore, it is decisive to see the unity of this process that does not exclude, but includes all other processes. This means that the judge's right does not play off against and constitution or against axiomatisation of law in the code (Esser, Dworkin).

From this follows a dynamical or process-logical order of the system of foundation:

- The starting point is the logical-grammatical interpretation, which represents, as it were, the surface of the system of foundation. It refers to the terminologically fixed sphere of wording in the code and includes the formal-logical coherence or consistence of the whole system.
- Furthermore, the historical, genetic interpretation has to be considered. It follows from the purpose of the law which can be deducted from history and its material.
- Finally, the systematic interpretation must make explicit not only the formal, but also the functional coherence of the whole system of legal order.

Juristic argumentation is sufficient and thus the basis of decision in a single case, if – and only if – this classical canon of interpretation is exhausted (See Raisch, 1988, p. 81). Particularly, this applies to systematic interpretation which does not only mean functional legal security, but refers to the justice of the whole system and thus makes itself amenable to the trans-positive legal-ethical principles.

3.3.4. *The Idea of Justice*

My starting point is Dreier's conception of juristic theory of argumentation as the continuation of Kant's law of reason. Likewise, Alexy – the main representative of juristic theory of argumentation in German speaking countries – assumed that with respect to its basic assertions this theory is to be called "Kantian". Contrasting legal positivism (which prohibits the recourse to an extra-juristic foundation of positive law) this theory claims that juristic argumentation is to be seen only within a general, practical theory of argumentation, i.e. ethics or moral doctrine, and its fundamental principles in all their varieties were anticipated by Kant's practical philosophy (Alexy, 1983, p. 153).

Dreier (1986) as well made explicit the argumentation-theoretical conception of justice by means of Kant's distinction between legal concept and legal idea.

The meaning of the Kantian legal concept is still given by its distinction between true, reasonable or just law and empirical or positive law, and by its assertion that there is a constitutive relation between law as it ought to be and law as it is in any historical situation. Hence, it goes beyond the controversy between natural law and legal positivism in all their possible kinds.

The distinction between empirical and ideal law as well as their relation is a question of practical reason or, as also Kant said, "reasonable will". The concretisation of the purely theoretical idea of law, which has been repeatedly postulated ever since Plato, remains to be a task that can never be accomplished. Plato locating the idea of justice in a supernatural sphere was himself aware of this problem. However, the Aristotelian doctrine of practical truth has brought the idea of justice from Plato's heaven back to the earth. So this idea lost, as we have known since Kant, its absolute realization.

For Kant it was an insight of practical reason that not everybody is able to arbitrarily see the given legal order as "doubtful law" (*ius controversum*). According to him, it was denied even to a juristic "scribe" to reason along these lines. As an official person, the jurist has to refer to "the code, which is publicly given and sanctioned by the highest authorities" (See Dreier, 1986, p. 10). Behind this formulation, however, is not a rigorous legal positivism (as it has been allegedly advocated in the spirit of Kant), but the conviction that law, as it actually exists, possesses a certain degree of rationality which must always be proved and secured. A purely empirical, i.e. positivist doctrine of law was for Kant, as I have already mentioned a "brainless" matter.

Each deeper foundation of law as "just" law must refer, after all, to the notion of *freedom*. Like Hegel, Kant saw the development of positive law basically as progress of the mind conscious of freedom, and argued that there is reason in the history of humankind.

Using freedom includes an essential distinction, which is relevant to the relation between ideal and positive law:

- The inner use of freedom refers to the motivation of an action according to the principle of virtue. It leads to laws of virtue that are no compulsory norms, but are followed by any single human being with resort to his/her inner motivation.

- The outer use of freedom refers to laws, which are, at least partly, commanded: In the natural condition of humans by the self-regulated society, in the civil condition by the state.

Crucial to this distinction is that a significant part of the outer laws is realized through inner legislation. Kant spoke of "natural laws", though he should better have spoken of "laws of reason" (See Dreier, 1986, p. 12). Certainly, both the classical natural law and the Kantian "law of reason" are, to certain degree, incorporated into positive law that can be enforced by the state.

This incorporation (= transformation), however, depends on a principle condition: The generalization of inner motives or subjective maxims.

The feature of definition for such natural laws, which can be transformed into positive law and were even explicitly characterized by Kant as "innate", is according to Dreier's reformulation "the compatibility of freedom of one person with the freedom of all the others" (Dreier, 1986, p. 13).

This formulation is tautological only with regard to the highest principle, which is without any content, the categorical imperative in its juristic use or the so-called "general principle of law" (Kant), which is nothing else but a normative formulation of (Kant's) concept of law (Dreier, 1986, p. 14). What is important is that Kant gained all the other statements of natural law through accumulation of general anthropological or – in today's sense – general *sociological* assumptions of empirical kind. But "the more concrete these assumptions are, the more dubious become the statements derived from them in connection to the principle of generalization and in the absence of secure empirical findings" (Dreier1986, p. 14). Such dubious statements are mixed with incomplete empirical arguments and permanently call for foundation, which, again and again, has to take resort to deeper groundings of reasonable law. At the end, the model of a deeper foundation refers to a process of optimisation which gives the principle of generalization – completed by "corollary principles" (Dreier) like the principles of consensus, changing roles, and – a regulative and heuristic function. This means that here we have principles which oblige us to achieve the best possible completeness, i.e. the highest possible degree of freedom with regard to positive rules and commands (juristic rules in the narrower sense). Hence the only legal constitution, which grows from this process of optimising the development of a sense for freedom is, according to Kant, "the pure republic".

Yet another distinction has to be made here: the distinction between principles and rules, which have been mentioned previously (See above 2.2.3.). Systematically the distinction means the following:

- The concept of rule is to be characterized by the fact that whenever it is put into practice, a command, a prohibition or a permission follow.
- Legal principles, however, do not always bring solutions of concrete legal problems. They are general normative maxims, which can be linked to values that can be put into practice only gradually.

Therefore, also from the point of view of contemporary theory of argumentation principles are closely related to values (Alexy, 1985, p. 125): The gradual realization of principles corresponds to the gradual realization of values. Dreier formulated this connection in the sense of Kant as follows:

> Practical reason in general can axiologically be regarded as capacity of practical ideas that can be deontologically transformed into practical principles of commands of maximisation or optimisation (Dreier1986, p. 17).

3.3.5. *Sufficient Saturation and Deeper Foundation of Juristic Argumentation*

Now we have approached the central problem of legal philosophy as theory of practical reason. Unlike legal positivism, this theory does not include any debate concerning the separation of law from morality

The conceptual and logical distinction between these two spheres has in fact to be accepted, but the main task is to show a constitutive relation between law and morality. In the words of the argumentation-theoretical conception of justice: Sufficient saturation of juristic argumentation through deeper legal-ethical principles that follow from the notion of justice as idea of practical reason.

Saturation of arguments is a claim, which can be found already – and explicitly – in Aristotelian philosophy and has been regarded as consequent deduction from sufficient reason.

However, the special problem in our case is the gap between the two spheres of is and ought in legal philosophy and morality. This gap cannot be simply bridged over by deduction, but one has to find other ways to build such a bridge.

The "deeper" foundation has, as it has been frequently recognized in the theory of argumentation, not the usual and non-problematic form of the deductive syllogism. Yet it is not only intuitive, speculative or even irrational one. One rather has to extend the concept of rational argumentation beyond mere deduction – the "normal case" of logical inference. This can be done in two ways:

- By "empirical evidence" (inductive inference), which is logically not valid as conclusion from the special case to a general principle, but is empirically true if one disposes of experience (Kant).
- By "practical conclusions" or substantial argumentation (Toulmin) which are not logical operations (because they go through different levels), but rather categorical transformations which rely on the persuasive power of the arguments.

In the process of juristic argumentation both ways are used in addition to deduction and subsumption. This process has to make clear the empirical fact by means of "juristic induction" and further empirical, descriptive statements considering historical, sociological, and, if necessary, also biological issues; moreover, the process has to reveal the values which play there role in each juristic decision. Thus, the person who is directly afflicted by the judicial verdict should be able to rationally understand the reasons for the particular decision. He or she should recognize that rational argumentation – and not an arbitrary act – has determined the verdict (quoted after Raisch, 1988, p. 22), and that "practical reason" has been used. However, the use of practical reason follows some rules, which can be rationally reconstructed. They are structures of deeper foundation and have been elaborated on the basis of Pecznik's and Alexy's argumentation theory with reference to Toulmin's general theory of argumentation (Toulmin, 1958) and his particular presentation of ethical arguments (Toulmin, 1950).

Toulmin's own starting point is Aristotle's conception of the capacity of truth inherent in practical, i. e. normative moral statements. However, his thesis is that apart from deductive conclusion (which he ascribed to Aristotle) there are special rules for moral inference to make possible the transition from facts to norms. Such rules should help to distinguish not only between valid and invalid moral arguments, but also between true and false moral statements. According to Toulmin a moral statement is true, if one has good reasons for it. Giving good reasons is in his opinion not identical with a resort to a general statement from which the special moral statement could be analytically derived. Practical arguing rather means a non-analytical specification of synthetic reasons. These are not only synthetic empirical statements made *a posteriori* (in jurisprudence statements of facts which are always synthetic), but general and substantial moral reasons that are located at a deeper level. Hence Toulmin distinguished between two levels and types of moral arguments:

At the first level is the justification of individual action. This corresponds to the deontological form of argumentation by which an action is justified, because a particular moral rule in a given society requires this action.

At the second level moral rules have to be justified which corresponds to the second, the teleological form of argumentation. In this case the arguments refer to the consequences of an action. They come into play only when two rules are in conflict or when none of them can be applied. This is a general discourse-theoretical consideration, and if one applies the two-level foundation to juristic argumentation, then the question is, where in this discussion – which is basically regulated by the positive legal system – the deeper foundation becomes necessary. The answer can be an argument of principles or an argument of injustice (See Dreier, 1986, Bydlinski, 1988).

The latter is the classical formulation of the resistance to authority. It means that positive law can develop into injustice, so that one is not just permitted, but obliged to resist. This is a contradiction to legal positivism which has always included the assertion that we have no right to oppose positive law, no matter on what it is based: on natural law, on the Bible, on conscience, on traditional law or on revolutionary ideas. The historical background of this view is, according to Kriele (1979, pp. 111-113) the idea of the sovereignty of the modern state which emerged in the 16th century under the impression of the religiously motivated civil wars.

The conception of the sovereign state, which is best articulated in Hobbes' *Leviathan*, is directed, in first instance, against rebellion and anarchy, and includes a pleading for order, obedience, and peace within a state. The argument for peace was also the basis for Kant's apparent legal positivism (Radbruch). In the case of a conflict between justice and peace, peace has to get priority for the idea of "eternal peace" is, in Kant's view, not only a philanthropic, but also a legal principle:

> For the legal condition within a state this means that [the condition] has in fact to be just. However, as long as it is not, positive law requires absolute obedience in order to guarantee, for reason's sake, the concretisation of the idea of secure law (Dreier, 1986, p. 21).

Therefore, Kant had his problems with the resistance to state authority (Dreier) but this was not his undoing. His comment on the French Revolution, which he explicitly regarded as the "evolution of the natural-legal constitution", makes clear that he decided to vote for the legal-political reform that can be necessitated by reason. Notwithstanding the congruency between positive and natural law (or law of reason, respectively), Kant's integrative concept of law denies that an "arbitrary law" implies any legal quality. Let's consider his formula "man and purpose" in the categorical imperative: Act in such a way that you need humanity, in your own person as well as in all the others, at all times as purpose and never as mere means (After Dreier, 1981, p. 291).

This formula has been used in the 20th century as a basis for the argument about injustice (Radbruch, 1963, p. 345, Welzel, 1962, p. 170, pp. 239-240).

The argument excludes certain kinds of norms, namely those that contradict "fundamental principles of justice so evidently that a judge who would apply them and recognize their legal consequences would prefer injustice to justice" (See Dreier, 1981, p. 33).

Here we have now a first "open problem" in the system of positive law. We clearly see the constitutive context of morality and law. Obviously, the question is, how to evaluate from a moral point of view the "legal-ethical quality" of positive legal principles.[7]

The second problem is the so-called "argument of principle". The positive legal system is – for reasons of legal security – a closed system. But this argument implies in its structural-theoretical version an almost classical foundation of Kant's theory of the legal concept.

The argument principles means that in all developed legal systems there are principles that, by virtue of its structure, go beyond the positivist legal concept. Dreier (1981: 28ff) has shown the substance of this argument on the basis of three theses:

1. The thesis of incorporation says that legal-ethical or, expressed in traditional terminology, principles of natural law, are incorporated in all developed legal systems.
2. The thesis of openness means that through this incorporation in developed legal systems positive law is, to a high degree, open to legal-ethical argumentation.
3. The thesis of optimization, finally implies that these legal-ethical principles are to be regarded as basic principles of law which impose legal obligation upon the legal systems incorporating them.

The first two theses can be accepted by advocates of each type of legal positivism claiming the separation of law from morality. With respect to the thesis of openness, the two most prominent representatives of legal positivism, Kelsen and Hart, have explicitly made this point, but in addition maintained that positive law is incomplete in this context. The thesis of optimisation, however, formulates the obligation of the judge to law in this "open sphere" which is not compatible with the separation of law from morality. The thesis signifies that the judge does not need to – and should not – have resort to extra-legal standards since all standards that are relevant to his/her decision are given by legally valid principles (Dreier, 1986, pp. 28-30). Such a presentation of the optimisation thesis makes modern theoretical formulation of the argument of principles identical to Kant's classical strong version of the argument. Kant called for an absolute and necessary congruity

[7] In accordance to Dreier's argument of injustice see also Bydlinski, 1988, p. 26.

of positive law and transformed natural law into a law of practical reason. Applied to the model of deeper foundation this clearly shows that each deeper juristic foundation doesn't leave the frames of the positive legal system to enter a vague and non-secured sphere of morality but it rather penetrates further the legal system, the latter providing – for the sake of legal security – the legal-ethical assessment standard.

According to Dreier such a theory of principles replaces (in the sense of Kant) the non-completeness thesis with the thesis of completeness thus reformulating the thesis of openness into a thesis of closeness. However this doesn't represent a turn-back to the completeness dogma of the old concept jurisprudence thus making legal development impossible and reducing the judge to a decision-making machine. In this case the notion of closeness doesn't imply the formal static and axiomatic-deductive closeness of a system of statements. Rather it refers to a dynamic and operational closing which enriches the positive legal system with essential explicit concretisations of the implicitly pre-supposed legal-ethical principles. Consequently legal development should be perceived as an ever-lasting self-construction of law.

3.4. *Rational Self-Construction of Law*

The phrase "rational self-construction" should be seen for what it is – namely a deliberate provocation against the self-organisational paradigm and the sociological system theory opposing self-construction to rationality.

On the other hand the basic pre-requisite for a legal philosophy as a theory of practical reason is the assumption that the rational argumentation of notions such as "values", "principles", "basic norms" or "just law" resembles very much action. In objection Luhmann claimed that to stick to the myth of reason is inadequate considering the complexity of modern society. The much richer in possibilities world, which could be completely clarified neither by theoretical nor by practical reason forbids such a constriction to reason: "The selection manner of the experience differs too much from the selection manner of action" (Luhmann, 1975).

Actually Luhmann's argument of complexity or hyper-complexity could be compared to Friedrich von Hayek's claim of "overestimation of reason" in his theory of complex systems. On cannot but agree with the idea that complex systems come to life (although not intentionally planned and in this sense not really completely clear) through human activities. Nevertheless this is hardly an argument for the lack of reason behind the process, the latter being too easily reduced to "habit" and in the case of legal questions – to blind habitual law by Hayek (in the spirit of Hume).

The reasonability of the world, of history, or of the whole – according to the respective classical terminology adopted since Kant and Hegel – is quite an acceptable empirical and even natural-scientific (biological) justification. The human being is – his individuality notwithstanding – the most-dependent on social cooperation living being (See also Oeser, 1987, pp. 126-127). This fundamental dependency of the human individual on the community performs itself in the constitutional state as a voluntary commitment to law, the latter being created not by an individual, sovereign ruler or by a particular authoritative body but rather being the result of collective reason. This view of law is best illustrated by Kant: The legislator couldn't rule about the people if the people cannot rule about itself (Quoted after Dreier, 1981, p. 14).

What follows is a presentation of a self-construction theory of law opposed to the classical imperative theory of law advocated by Bentham and Austin. For, strictly speaking, there is no higher authority, which could create laws independently from the people not to mention completely arbitrary and sovereign. In this sense one cannot but agree with Dreier who revising Kant's definition of law determines the latter as the sum-total of acts of law which connects the freedom of the individual with the freedom of the many (Dreier, 1981, p. 15).

This definition comprises both natural law and positive law presupposing a constitutive connection between them, which reveals itself as a dynamical and developmental process of constant self-construction. The process of self-construction of law is best represented through the connection of two classical basic concepts of legal theory, the one representing the static-constructivist side and the other – the dynamic-procedural side. These are namely the traditional concept jurisprudence with the fundamental notion of juristic construction on the one hand, and the originating from the hermeneutical tradition self-referential applicative circle, on the other.

Both these conceptions could be very well connected to the juristic argumentation theory. As it will be further shown, the connection between the argumentation theory and the concept jurisprudence was at least assumed by Dreier. With reference to the applicative circle Alexy explicitly stated that the external as well as the internal justification is compatible with the already classical model of Engisch – the constantly changing perspective between the state of facts and the norm: In the case of internal justification one can (with the help of arguments valid for the juristic discourse) argue separately the state of facts with its peculiarities and the specificity of the norm. Whereas in the case of external justification one tries to bridge over the gap between the norm and the description of facts. (Alexy 1983, p. 282)

Alexy however supported the dogma of the old philosophy of science about the demarcation between the process of discovery and the process of justification. Although he correctly defined Kriele's model of finding and verification of norm hypotheses as a process of discovery, Alexy's attitude towards the long ago revised (see Raisch, 1988, pp. 26-27, Maasch. 1986, pp. 361-362) by the philosophy of science dogma reflects the linear structure of argumentation theory. This linear structure could be overcome through the incorporation of the argumentation theory (restricted only to the process of justification) in the cycle of discovery and retroactive justification.

The applicative circle as a hermeneutical circle embodies the traditional canons interpretation. In the juristic discourse theory it brings the process of discovery and the process of justification together in the cyclic model of a dynamic self-referential system.

This model could be used to clarify yet another problem of traditional juristic methodology. For, the meaning of the concept of systematic interpretation in the conventional canons of hermeneutics (traced back not only to Savigny), has remained vague (see Raisch, 1988, pp. 26-27, Maasch,. 1986, pp. 361-362) and ambiguous. Thus the systematic interpretation is either identified with the logical coherence of the legal system, or with the teleological interpretation interested in the final goal of the whole system of legal regulation, or with the integration of logical-grammatical interpretation and the historical one. Here the model of the applicative circle could be used in the following way: Systematic interpretation should be understood as the functional coherence of the whole argumentation cycle which leads from case to norm and then backwards – from norm to the case again. On the one hand functional coherence has to be understood in its trivial sense – as a coherent symmetrical functioning of the legal system as a whole in all its parts: e.g. what is to the right in one part should be to the right in the other parts as well. On the other hand however the concept of function in its initial mathematical and logical sense means "a state of non-saturation". This state of any function (according to Frege each function f(x) is characterised by a state of non-saturation; hence it has to be constantly 'filled' with an argument (=x)), requires an argumentative saturation in the spirit of juristic argumentation theory. Compared to mathematics and logic however here the process aims at the best possible level of saturation and not necessarily at the complete realisation of arguments i.e. the foundation should encompass all reasonable arguments which are of importance to the relatively (according to the system of positive law) correct ruling (See Alexy, 1983, p. 293, Raisch, 1988, p. 23). With his popular model of the ever changing perspective between norm and case, Engisch had already hinted at such best possible saturation as a prerequisite for the best possible decision. The model however represents nothing else but the applicative circle in jurisprudence. This methodological cycle (also as a developmental process) reveals in both directions – empiric and systematic – a double meaning: On the one hand it reflects the growth of solution efficiency on the level of single cases, on the other – the growth of the legal system itself. In terms of systems theory this is a positive expansion and not only a negative feedback in the sense of a reactive consolidation.[8] With reference to juristic methodology this means that prejudication could replace missing norms. However this practice of replacement would lead to a retroactive consolidation of juristic decisions (forbidden in jurisprudence) thus hindering the development of the legal system. Yet the progress of the legal system is based on the transformation of legal-ethical evaluations into normative principles for juristic decision-making.

The model of profound foundation doesn't presuppose an open legal system permitting loopholes and vague moral evaluations. Rather it requires a closeness of the system thus changing the form of concept jurisprudence. In this sense Dreier correctly assumed that the defamed basic hypotheses of concept jurisprudence were open at least to a principle-theoretical interpretation.

This would be an appropriate place to elaborate more systematically the relationship between values and principles (already mentioned by Alexy). By saying that the gradual realisation of principles corresponds to the gradual realisation of values one means that the respective moral evaluations are to be realised through their transformation into principles of a positive legal system. The transformation has proved to be the only way to penetrate a normatively closed system such as the legal system construed according to the good old tradition of German concept jurisprudence. A particular legal order is valid only if it contains those moral evaluations which transformed into legal-theoretical principles serve positive law as basic norms. It is however not necessary that they are explicitly present or wholly and precisely formulated – they should be only non-conditionally valid. These basic norms prescribe (and not hypothetically but rather categorically – in the spirit of the categorical imperative functioning as principle of generalisation) whether an action is to take place or not.

The non-conditionality of these principles is however necessarily connected to a higher grade of uncertainty. The latter admits linguistic formulations – defined already by Kant as principles of human dignity, freedom, and equality –,which differ from one another in the respective constitutions (See Dreier, 1981, p. 23). The well developed legal systems boast of high quality an important part of which is the differentiation in the

[8] On the problem of hermeneutical spiral see e.g., Schroth, 1977, Riedl 1985, p. 276.

formulation of their principles. Yet the quality of a legal order is to be practically demonstrated through its efficiency. The latter should not be understood only and primarily as a legal-sociological expression of the social acceptance of a judge's ruling. Rather it reflects the immanent capacity of the legal system to provide arguments for a concrete and adequate to the case decision. This capacity shows itself mainly by non-trivial and complex juristically relevant facts of the case, which could best reveal the problem-solution deficiency of a legal system. Then it has always been the just ruling of a complex case to determine the quality of the legal system.

On the other hand the quality of the legal system is founded in the implicitly pre-supposed by the system quality of legal-ethical principles. Respectively one could speak of a constitutive connexion between implicitly pre-supposed legal-ethical principles, on the one hand and explicitly formulated in the positive legal system normative ones on the other which together reflect the systematic relation between morality and law. This connexion could be best expressed on the basis of the (in this context often applied (Teubner, 1986)) model of self-organisation paradigm interpreted however mainly in the terms of legal sociology. Yet the sociological interpretation of legal principles leads always to a constitutional obscurity (Stern, 1973, p. 14, Dreier, 1981, p. 158). To avoid it the self-organisation paradigm here will be transformed into a legal-philosophical self-construction paradigm. In the sense of a strict restoration of concept jurisprudence (however without its exaggerations and lack of formalism) the self-construction paradigm could be expressed as follows: Each fully developed legal order construes a normatively closed system possessing its own system-relative environment, the so-called legal reality. The latter represents a superimposition of natural and social realities of the human being. One would never wholly (in the spirit of Kant) get to know it, nevertheless the legal reality is basically accessible to cognitive processes. Such a cognitive openness is characteristic especially for a legal reality related to a particular normative legal system, i.e. the legal reality reflects only this part of juristically relevant facts, which is covered by the respective legal system. Nevertheless the system of legal concepts is constrained exclusively by a regulating mechanism.

This regulating mechanism represents nothing else but the fact that the legal system cognitively opens itself for particular facts of life and practical assessments and then closes itself again through the act of transformation of those assessments into positive norms and rules which thus become a part of the system. The same procedure is valid also for more general and profounder legal-ethical principles, which come to life through the evolutionary development of moral systems. Ethics as a philosophical discipline has always systematically tried to justify those principles which are however to be concretely realised only with the help of their transformation into positive law.

The system of positive law is an autonomous system independent from state authority and politics due to the fact that its basic norms rest on profound and legal-ethical principles. No legislator could violate those principles without provoking the resistance of natural law thus preparing his/her own fall.

This demonstrates the constitutional and active role of positive law in the development of the human moral system, which has always been an object of philosophical speculations but not of practical deeds. Then a real development of legal-ethical principles and thereby of human morality could be achieved only if at all through the coercive system of positive law.

REFERENCES

Aarnio, A. (1979) *Denkweisen der Rechtswissenschaft*, Vienna-New York.

Alexy, R. (1983) *Theorie der juristischen Argumentation. Die Theorie des rationalen Diskurses als Theorie der juristischen Begründung*, Frankfurt a. M.

Alexy, R. (1985) *Theorie der Grundrechte*, Baden-Baden.

Amira K. v. (1891) *Tierstrafen und Tierprozesse*, Innsbruck.

Ampére, A. M. (1834-43) *Essai sur la philosophie des sciences, ou exposition analytique d'une classification naturelle de toutes les connaissances humaines*, Paris.

Apel, K.O. (1955) Das Verstehen. *Archiv für Begriffsgeschichte*, I, Bonn.

Aristotle, *Organon* [Translated by E. Rolfes], vol. 1.

Ballweg, O. (1970) *Rechtswissenschaft und Jurisprudenz*, Basel.

Beer, S. (1966) *Decision and Control*, New York-London.

Betti, E. (1967) Allgemeine Auslegungslehre als Methodik der Geisteswissenschaften, Tübingen.

Bühl, W. L. (Ed) (1975) *Verstehende Soziologie*, Munich.

Bydlinski, F. (1988) *Fundamentale Rechtsgrundsätze*, Vienna-New York.

Canaris, C. W. (1964) *Die Feststellung von Lücken im Gesetz*, Berlin.

Canaris, C. W. (1969) *Systemdenken und Systembegriff in der Jurisprudenz, entwickelt am Beispiel des deutschen Privatrechts*, Berlin.

Carnap, R. (1974) *Einführung in die Philosophie der Naturwissenschaft*, Munich.

Childe, G. (1975) *Soziale Evolution*, Frankfurt.

Coing, H. (1953) Zur Methodik der republikanischen Jurisprudenz, *Studi Arangio-Ruiz*, I, Napoli.

Coing, H. (1972) *Juristische Methodenlehre*, Berlin-New York.

Constantinesco L.J. (1972) *Rechtsvergleichung*, Cologne-Berlin-Bonn.

Darwin, Ch. (1975) *Die Abstammung des Menschen und die geschlechtliche Zuchtwahl.* [Translated by J. V. von Carus], Stuttgart.

Diemer, A. (1970) *Der Wissenschaftsbegriff. Historische und systematische Untersuchungen*, Meisenheim.

Diemer, A. (1971) *Der Methoden- und Theorienpluralismus in den Wissenschaften*, Meisenheim.

Diemer, A. (1975) Der Wissenschaftsbegriff in den Natur- und Geisteswissenschaften. *Studia Leibnitiana*, 5, Wiesbaden.

Dilthey, W. *Gesammelte Werke,* vol. 5.

Dreier, R. (1981) *Recht– Moral – Ideologie*, Frankfurt a. M.

Dreier, R. (1986) Rechtsbegriff und Rechtsidee. Kants Rechtsbegriff und seine Bedeutung für die gegenwärtige Diskussion. *Würzburger Vorträge zur Rechtsphilosophie, Rechtstheorie und Rechtssoziologie*, Frankfurt a. M.

Eder, K. (1980) *Die Entstehung staatlich organisierter Gesellschaften*, Frankfurt.

Eder, K. Zur Rationalisierungsproblematik des modernen Rechts. *Soziale Welt*, 2.

Emge, C.A. (1916) *Über das Grunddogma des rechtsphilosophischen Relativismus*, Berlin.

Eneccerus-Nipperdey (1959) *Allgemeiner Teil des Bürgerlichen Rechts.*

Engisch, K. (1963) *Logische Studien zur Gesetzanwendung*, Heidelberg.

Engisch, K. (1968) *Die Idee der Konkretisierung in Recht und Rechtswissenschaft unserer Zeit*, Heidelberg.

Engisch, K. (1971) *Auf der Suche nach der Gerechtigkeit. Hauptthemen der Rechtsphilosophie,* Munich.

Engisch K. (1977) *Einführung in das juristische Denken*, Berlin-Cologne- Mainz.

Esser, J. (1956) *Grundsatz und Norm in der richterlichen Fortbildung des Privatrechts*, Tübingen.
Esser, J. (1972) *Vorverständnis und Methodenwahl in der Rechtsfindung*, Frankfurt a. M.
Fikentscher, W. (1975-77) *Methoden des Rechts in vergleichender Darstellung*, Tübingen.
Flume, (1975) *Allgemeiner Teil des Bürgerlichen Rechts*, vol. 2.
Forsthoff, (1959) *Festschrift für C. Schmitt.*
Freeman, D. (1974) The Evolutionary Theories of Charles Darwin and Herbert Spencer. *Current Anthropology*, 15, pp 211 – 237.
Gadamer, H .G. (1965) *Wahrheit und Methode*, Tübingen.
Galileo, G. (1891) *Dialog über die beiden hauptsächlichen Weltsysteme, das Ptolemäische und das Kopernikanische*. [Translated by E. v. Strauss], Leipzig.
Glansdorff, P., I. Prigogine (1971) *Thermodynamic Theory of Structure, Stability and Fluctuations*, New York.
Goodhart, A. L. (1934) *Precedents in English and Continental Law.*
Grimm, D. v. (Ed.) (1976) *Rechtswissenschaft und Nachbarwissenschaften*, Munich.
Gschnitzer, F. (1946) *ÖJZ.*
Habermas, J. (1968) Erkenntnis und Interesse. *Technik und Wissenschaft als Ideologie*, Frankfurt.
Habermas, J. (1976) Zum Thema: Geschichte und Evolution. *Geschichte und Gesellschaft*, 3.
Hamson, (1950) Droit comparé et entseignement du droit. *RIDC.*
Hart, H. L. A. (1961) *The Concept of Law*, Oxford. [Der Begriff des Rechts, Frankfurt a.M., 1973]
Hart L. A. (1971) *Recht und Moral* (translated and edited by N. Hoerster), Göttingen.
Hassemer, W. (1968) *Tatbestand und Typus*, Cologne-Berlin-Bonn-Munich.
Hayek, F.v. (1983) Die überschätzte Vernunft. In: Riedl R. J., F. Kreuzer (Eds.) *Evolution und Menschenbild*, Hamburg.
Heck, Ph. (1932) *Begriffsbildung und Interessenjurisprudenz*, Tübingen.
Heck, Ph. (1912) *Das Problem der Rechtsgewinnung.* Rektoratsrede. [Edited by J. v. Esser (1968)].
Hempel, C. G., P. Oppenheim (1936) *Der Typusbegriff im Lichte der neuen Logik*, Leiden.
Hempel, C. G. (1977) *Aspekte wissenschaftlicher Erklärung*, Berlin-New York.
Hobbes, Th. (1949) *Lehre vom Menschen, Lehre vom Bürger*. [Translated by M. v. Frischeisen-Köhler], Leipzig.
Hoerster, N. (Ed.) (1979) *Klassische Texte der Staatsphilosophie*, München.
Hoerster, N. (1982) Rechtsethik ohne Metaphysik. *Juristen-Zeitung*, pp. 264 – 272.
Horak, F. *Rationes Decidendi. Entscheidungsbegründungen bei den älteren römischen Juristen bis Labeo*, Aalen.
Hruschka, J. (1972) *Das Verstehen von Rechtstexten. Zur hermeneutischen Transpositivität des Rechts.*
Huxley, J. (1942) *Evolution, the modern synthesis*, New York.
Ihering, R. (1852 – 63) *Geist des römischen Rechts auf den verschiedenen Stufen seiner Entwicklung.* Basel o.J.
Ihering, R. (1893) *Der Zweck im Recht*, Leipzig.
Inciarte F. (1974) Theoretische und praktische Wahrheit. In: Riedel M. (Ed.) *Rehabilitierung der praktischen Philosophie*, Freiburg.
Jahr, G. W. Maihofer (Eds.) (1971) *Rechtstheorie*, Frankfurt a. M.
Jörs, P. W. Kunkel (1949) *Römisches Privatrecht.*
Kants Werke (1968), Akademie Textausgabe, Berlin.
Kaser, M. (1962) Zur Methode der römischen Rechtsfindung. *Nachricht der Akademie der Wissenschaften, Phil. Hist. Kl.*, 2, Göttingen.
Kaser, M. (1967) Der Römische Anteil am deutschen bürgerlichen Recht. *JUS.*

Kaufmann, A. (1962) Gesetz und Recht. In: *Existenz und Ordnung.* Festschrift für E. Wolf, Frankfurt a. M.
Kaufmann, A. (1965) *Analogie und "Natur der Sache". Zugleich ein Beitrag zur Lehre vom Typus,* Karlsruhe.
Kaufmann, A. (1971) Die Geschichtlichkeit des Rechts im Licht der Hermeneutik. In: Kaufmann, A. (Ed.) *Rechtstheorie, Ansätze zu einem kritischen Rechtsverständnis,* Karlsruhe.
Kaufmann, A. (1984) *Beiträge zur Juristischen Hermeneutik sowie weitere rechtsphilosophische Abhandlungen,* Cologne-Berlin-Bonn-Munich.
Kaufmann, A. (1984a) *Rechtsphilosophie im Wandel,* Cologne-Berlin-Bonn-Munich.
Kaufmann, A., W. Hassemer (Eds.) (1977) *Einführung in die Rechtsphilosophie und Rechtstheorie der Gegenwart,* Heidelberg-Karlsruhe.
Kaulbach, F. (1968) *Philosophie der Beschreibung,* Cologne-Graz.
Kaulbach, F. (1979) Rechtsphilosophie und Rechtstheorie bei Kant. In: Heintel E. (Ed.) *Philosophische Elemente der Tradition des politischen Denkens,* Vienna-Munich.
Kedrow, B. W. (1976) *Klassifizierung der Wissenschaften,* Moscow.
Kelsen, H. (1960) *Reine Rechtslehre,* Vienna.
Klug, U. (1966) *Juristische Logik,* Berlin-Heidelberg-New York.
Koschaker, P. (1958) *Europa und das römische Recht.*
Kraft, V. (1960) *Erkenntnislehre,* Vienna.
Kraft, V. (1973) *Die Grundformen wissenschaftlicher Methoden,* Vienna.
Krawietz, W, R. Alexy (Eds.) (1983) *Metatheorie der Juristischen Argumentation,* Berlin.
Kriele, M. (1967) *Theorie der Rechtsgewinnung,* Berlin.
Kriele, M. (1979) *Recht und praktische Vernunft,* Göttingen.
Larenz, K. (1979) *Methodenlehre der Rechtswissenschaft,* Berlin-Heidelberg-New York.
Leenen, D. (1971) *Typus und Rechtsfindung,* Berlin.
Lenk, H. (1971) *Neue Aspekte der Wissenschaftstheorie,* Braunschweig.
Liebs, D. (1975) *Römisches Recht.*
Locke, J. (1967) *Zwei Abhandlungen über die Regierung,* Frankfurt a. M.
Lorenz, K. (1973) *Die Rückseite des Spiegels,* Munich.
Lorenz, K. (1974) *Analogy as a Source of Knowledge.* [Les prix Nobel 1973], The Nobel Foundation.
Lorenzen, K. (1968) *Methodisches Denken,* Frankfurt.
Luhmann, N. (1972) *Rechtssoziologie* I.
Luhmann, N. (1975) Evolution und Geschichte. *Geschichte und Gesellschaft,* 3.
Maasch, B. (1986) Auslegung von Normen – ein spezifisches Problem im Kartellrecht? *ZHR* 150, pp 354-365.
Maine, H. S. (1861) *Ancient Law. Its Connection with the Early History of Society and its Relation to Modern Ideals,* London.
Messer, J. (1950) *Das Naturrecht.*
Mezger, E. (1989) Die subjektiven Unrechtselemente. *Gerichtssaal.*
Mohr, H. (1987) *Natur und Moral. Ethik in der Biologie,* Darmstadt.
Morgan, L. H. (1871) *Ancient Society.*
Müller, F. (1976) *Juristische Methodik.*
Naucke, W. (1986) *Rechtsphilosophische Grundbegriffe,* Fernuniversität Hagen.
Nelson, L. (1949) *Rechtswissenschaft ohne Recht,* Göttingen-Hamburg.
Neumann, J. v., O. Morgenstern (1961) *Spieltheorie und wirtschaftliches Verhalten,* Würzburg.
Newton, I. (1872) *Mathematische Prinzipien der Naturlehre.* [Edited by J. Ph. V. Wolfers], Berlin.

Nörr, D. (1972) *Divisio und partitio, Bemerkungen zur römischen Rechtsquellenlehre und zur antiken Wissenschaftstheorie*, Berlin.
Oeser, E.(1976) *Wissenschaft und Information*, Vienna-Munich.
Oeser, E. (1978) Die Dynamik wissenschaftlicher Begriffsformen. In: *Studien zur Klassifikation*, vol. 2.
Oeser, E. (1979) *Wissenschaftstheorie als Rekonstruktion der Wissenschaftsgeschichte*, Vienna-Munich.
Oeser, E. (1982) Komparative Wissenschaftstheorie. Jenseits von Szientistik und Hermeneutik. In: Nagl-Docekal H. (Ed.) *Überlieferung und Aufgabe*, Vienna.
Oeser, E. (1987) Theorie der soziokulturellen Evolution. *Wiener Studien zur Wissenschaftstheorie,* 1.
Oeser, E. (1987a) *Psychozoikum. Evolution und Mechanismus der* menschlichen Erkenntnisfähigkeit, Berlin-Hamburg.
Oeser, E. (1988) *Das Abenteuer der kollektiven Vernunft. Evolution und Involution der Wissenschaft*, Berlin-Hamburg.
Oeser, E. (1988a) *Wissenschaftspluralismus in der Medizin vom Standpunkt der komparativen Wissenschaftstheorie*, Nürnberg.
Oeser, E., F. Seitelberger (1988) *Gehirn, Bewusstsein und Erkenntnis*, Darmstadt.
Oppenheim, P. (1926) *Die natürliche Ordnung der Wissenschaften*, Jena.
Patzig, G. (1979) "Erklärung". In: Braun v., Rademacher (Eds.) *Wissenschaftstheoretisches Lexikon*, Cologne-Graz.
Peczník, A. (1979) Non-equivalent Transformations and the Law. *Rechtstheorie*, 1.
Peczník, A. (1983) *Grundlagen der juristischen Argumentation*, Vienna-New York.
Pollock, F. (1903) The History of Comparative Jurisprudence. *JoCLV*, 76.
Popper, K. R. (1973) *Logik der Forschung*, Tübingen.
Popper, K. R. (1973) *Objektive Erkenntrnis*.
Post, A. H. (1867) *Das Naturgesetz des Rechts*, Bremen-Halle.
Post, A. H. (1894-95) *Grundriss der ethnologischen Jurisprudenz*, Oldenburg-Leipzig.
Puchta, G. F. (1893) *Cursus der Institutionen*, Leipzig.
Radbruch, G. (1906) Über die Methode der Rechtsvergleichung. *MKSR*, 2.
Radbruch, G. (1938) *Klassenbegriffe und Ordnungsbegriffe im Rechtsdenken*.
Radbruch, G. (1946) *Geist des englischen Rechts*.
Radbruch G. (1963) *Rechtsphilosophie*, Stuttgart.
Raisch, P. (1973) *Unternehmensrecht*, 1.
Raisch, P. (1979) *JZ*.
Raisch, P. (1977) Ist mehr Wettbewerb möglich? *Hagener Seminarbeiträge zum Unternehmensrecht*, 1, Karlsruhe.
Raisch, P. (1988) *Vom Nutzen der überkommenen Auslegungskanones für die praktische Rechtsanwendung*, Heidelberg.
Rawls, H. (1975) *Eine Theorie der Gerechtigkeit*, Frankfurt a.M.
Reichenbach, H. (1938) *Experience and Prediction*, Chicago.
Rescher, N. (1977) Handlungsaspekte. In: Meggle G. (Ed.) *Analytische Handlungstheorie*, 1, Frankfurt a. M.
Rheinstein, M. (1974) *Einführung in die Rechtsvergleichung*.
Riedel, M. (Ed.) (1972-74) *Rehabilitierung der praktischen Philosophie*, Freiburg.
Riedl, R. (1980) *Biologie der Erkenntnis*, Hamburg-Berlin.
Riedl, R. (1985) *Die Spaltung des Weltbildes*, Berlin-Hamburg.
Riedl, R., F. Kreuzer (Eds.) (1983) *Evolution und Menschenbild*, Hamburg.
Riedl, R., F. M. Wuketits (Eds.) (1987) *Die evolutionäre Erkenntnistheorie*, Hamburg-Berlin.
Rottleuthner, H. (1976) Hermeneutik und Jurisprudenz. In: Koch H.-G. (Ed.) *Juristische Methodenlehre und analytische Philosophie*, Kronberg/Ts.

Rousseau, J. J. (1948) *Der Gesellschaftsvertrag*, Munich.
Ryffel, H. (1969) Grundprobleme der Rechts- und Staatsphilosophie. *Philosophische Anthropologie des Politischen*, Neuwied-Berlin.
Savigny, F. C. v. (1840) *Vom Beruf unserer Zeit für Gesetzgebung und Rechtswissenschaft*, Heidelberg.
Savign,y F. C. v. (1951) *Juristische Methodenlehre*. [Edited by v. G. Wesenberg], Stuttgart.
Scheler, M. (1954) *Der Formalismus in der Ethik und die materiale Wertethik*, Bern.
Schlosser, H. (1979) *Grundzüge der neueren Privatrechtsgeschichte*.
Schneider, H. P. (1972) Rechtstheorie ohne Recht? Zur Kritik des spekulativen Positivismus in der Jurisprudenz. In: *Mensch und Recht*. Festschrift für R. E. Wolf, Frankfurt a. M.
Schramm, Th. (1978) *Einführung in die Rechtsphilosophie*, Cologne.
Schroth, U. (1977) Probleme und Resultate der Hermeneutik. In: Kaufmann A., W. Hassemer (Eds.) *Einführung in die Rechtsphilosophie und Rechtstheorie der Gegenwart*, Heidelberg-Karlsruhe.
Schulz, F. (1934) *Prinzipien des römischen Rechts*, Munich-Leipzig.
Schulz, F. (1961) *Geschichte der römischen Rechtswissenschaft*, Weimar.
Schwanitz, H. J., E. Oeser (1985) Der Begriff "Exanthem" aus wissenschaftstheoretischer Sicht. *Der Hausarzt IV*, VI, 34.
Schwemmer, O. (1986) *Ethische Untersuchungen*, Frankfurt a. M.
Seidl, E. (1962) *Römische Rechtsgeschichte und römisches Zivilprozessrecht*, Cologne-Berlin-Bonn- Munich.
Semmel, M. (1984) *Die Unternehmung aus evolutionstheoretischer Sicht*, Berlin-Stuttgart.
Silving, H. (1958) A Plean for a Law of Interpretation. *University of Pennsylvania Law Review*.
Stegmüller, W. (1969) Probleme und Resultate der Wissenschaftstheorie und analytischen Philosophie. *Erklärung und Begründung*, Berlin-Heidelberg-New York.
Stegmüller, W. (1970) Probleme und Resultate der Wissenschaftstheorie und analytischen Philosophie. *Theorie und Erfahrung*, Berlin-Heidelberg-New York.
Stegmüller, W. (1973) Der sogenannte Zirkel des Verstehens. In: Hübner K., A. Menne (Eds.) *Natur und Geschichte. X. Deutscher Kongress für Philosophie*, Hamburg.
Stegmüller, W. (1974) *Personelle und statistische Wahrscheinlichkeit*, Vienna-Berlin-New York.
Stein, P. (1980) *Legal Evolution, The Story of an Idea*.
Steinwenter, A. (1951) Prolegomena zu einer Geschichte der Analogie. *Festschrift für F. Schulz*, II, Weimar.
Stephanitz ,D. v. (1970) *Exakte Wissenschaft und Recht*, Berlin.
Stern, K.(1973) *Verfassung und Verfassungsreform in der Bundesrepublik Deutschland*, Krefeld.
Stevenson, (1950) Comparative and Foreign Law. *American Law Schools.* Columbia L.
Stroux, J. (1949) *Römische Rechtswissenschaft und Rhetorik*, Potsdam.
Schulz, F. (1934) *Prinzipien des römischen Rechts.*
Tammelo, I. (1969) *Outlines of Modern Legal Logic.*
Teubner, G. (1986) Gesellschaftsordnung durch Gesellschaftslärm? Autopoietische Geschlossenheit als Problem für die Rechtssetzung. *EVI Working Paper* Nr 86/207, Badia Fiesolaner San Domenico.
Thur (1914) *Der Allgemeine Teil des Deutschen Bürgerlichen Rechts*, II.
Toulmin, S. E. (1950) *The Place of Reason in Ethics*, Cambridge.
Toulmin, S. E. (1958) *The Uses of Argument*, Cambridge.
Trimborn, H. (1950) Die Privatrechte und der Eingriff des Staates. In: Wolf E. (Ed.) *Deutsche Landesreferate zum III. Intern. Kongress für Rechtsvergleichung in London, 1950*, Tübingen.

Troeltsch, E. (1925) *Gesammelte Schriften IV,* Tübingen.
Vecchio, G. del (1937) *Der homo juridicus, Philosophia* II, Belgrade.
Viehweg, Th. (1963) *Topik und Jurisprudenz*, Munich.
Vollmer, G. (1975) *Evolutionäre Erkenntnistheorie,* Stuttgart.
Vinogradoff (1929) *Roman Law in Mediaeval Europe.*
Wahl, E. (1950) Methoden und Grenzen der Rechtsvereinheitlichung. *Beiträge zur Rechtsforschung,* Sonderveröffentlichung der Ges. f. ausl. u. intern. Privatrecht, Tübingen.
Wallace A. R. (1864) On the Origin of Human races deduced from the Theory of Natural Selection, *Anthropological Review.*
Weinkauf, H. (1970) Was heißt das: "Positivismus als juristische Strategie"? *Juristen-Zeitung.*
Welzel, H. (1958) *Die Naturrechtslehre Samuel Pufendorfs.*
Welzel, H. (1962) *Naturrecht und materiale Gerechtigkeit,* Göttingen.
Wesel, U. (1984) Bemerkungen zu einer evolutionistischen Theorie des Rechts. In: Nörr D., D. Simon (Eds.) *Gedächtnisschrift für Wolfgang Kunkel*, Frankfurt a. M.
Wieacker, F. (1958) *Gesetz und Richterkunst*, Karlsruhe.
Wieacker, F. (1961) *Vom römischen Recht.*
Wieacker, F. (1962) Die juristische Sekunde. *Festschrift für E. Wolf*, Frankfurt a. M.
Wieacker, F. (1963) *Notizen zur rechtshistorischen Hermeneutik*, Göttingen.
Wieacker, F. (1967) *Privatrechtsgeschichte der Neuzeit*, Göttingen.
Wieacker, F. (1981) Bemerkungen über Ihering und den Darwin, Ch.ismus. In: Altner G. (Ed.) *Der Darwinismus*, Darmstadt.
Wilhelm, W. (1958) *Zur juristischen Methodenlehre im 19. Jahrhundert*, Frankfurt a. M.
Wolf, E. (1951) *Große Rechtsdenker*, Tübingen.
Wolf, E. (1953) Fragwürdigkeit und Notwendigkeit der Rechtswissenschaft. *Freiburger Universitätsreden*, N.F. 15.
Wolf, E. (1964) *Das Problem der Naturrechtslehre.*
Wolf E. (1972) *Rechtsphilosophische Studien. Ausgewählte Schriften* I, Frankfurt.
Wolff, H. J. (1952) Die Grundlagen der griechischen Eherechts. *Tijdschrift voor Rechtsgeschiedenis* 20.
Wright, G. H. v. (1963) *The Varieties of Goodness*, London-New York.
Wright, G. H. v. (1963a) *Norm and Action*, London.
Wright, G. H. v. (1974) *Erklären und Verstehen*, Frankfurt.
Zemen, H. (1983) *Evolution des Rechts. Eine Vorstudie zu den Evolutionsprinzipien des Rechts auf anthropologischer Grundlage*, Vienna -New York.
Zippelius, R. (1971) *Juristische Methodenlehre*, Munich.
Zitelmann, (1879) *Irrtum und Rechtsgeschäft.*

THEORY AND DECISION LIBRARY

SERIES A: PHILOSOPHY AND METHODOLOGY OF THE SOCIAL SCIENCES

Editors: W. Leinfellner (*Vienna*) and G. Eberlein (*Munich*)

1. G. Zecha and P. Weingartner (eds.): *Conscience: An Interdisciplinary View*. Salzburg Colloquium on Ethics in the Sciences and Humanities (1984). 1987 ISBN 90-277-2452-0
2. R.W. Scholz: *Cognitive Strategies in Stochastic Thinking*. 1987 ISBN 90-277-2454-7
3. H. Nurmi: *Comparing Voting Systems*. 1987 ISBN 90-277-2600-0
4. M. Schmid and F.M. Wuketits (eds.): *Evolutionary Theory in Social Science*. 1987 ISBN 90-277-2612-4
5. C.L. Sheng: *A New Approach to Utilitarianism*. A Unified Utilitarian Theory and Its Application to Distributive Justice. 1991 ISBN 0-7923-1301-1
6. F. Forman: *The Metaphysics of Liberty*. 1989 ISBN 0-7923-0080-7
7. G. Bernard: *Principia Economica*. 1989 ISBN 0-7923-0186-2
8. J.W. Sutherland: *Towards a Strategic Management and Decision Technology*. Modern Approaches to Organizational Planning and Positioning. 1989 ISBN 0-7923-0245-1
9. C. Vlek and G. Cvetkovich (eds.): *Social Decision Methodology for Technological Projects*. 1989 ISBN 0-7923-0371-7
10. P. Hoyningen-Huene and F.M. Wuketits (eds.): *Reductionism and Systems Theory in the Life Sciences*. Some Problems and Perspectives. 1989 ISBN 0-7923-0375-X
11. K.G. Grunert and F. Ölander (eds.): *Understanding Economic Behaviour*. 1989 ISBN 0-7923-0482-9
12. G. Antonides: *The Lifetime of a Durable Good*. An Economic Psychological Approach. 1990 ISBN 0-7923-0574-4
13. G.M. von Furstenberg (ed.): *Acting under Uncertainty*. Multidisciplinary Conceptions. 1990 ISBN 0-7923-9063-6
14. R.L. Dukes: *Worlds Apart*. Collective Action in Simulated Agrarian and Industrial Societies. 1990 ISBN 0-7923-0620-1
15. S.O. Funtowicz and J.R. Ravetz: *Uncertainty and Quality in Science for Policy*. 1990 ISBN 0-7923-0799-2
16. J. Götschl (ed.): *Erwin Schrödinger's World View*. The Dynamics of Knowledge and Reality. 1992 ISBN 0-7923-1694-0
17. R. von Schomberg (ed.): *Science, Politics and Morality*. Scientific Uncertainty and Decision Making. 1993 ISBN 0-7923-1997-4
18. J. Baron: *Morality and Rational Choice*. 1993 ISBN 0-7923-2276-2
19. M. Allais and O. Hagen (eds.): *Cardinalism*. A Fundamental Approach. 1994 ISBN 0-7923-2398-X
20. E. Carlson: *Consequentialism Reconsidered*. 1995 ISBN 0-7923-3547-3
21. J. Götschl (ed.): *Revolutionary Changes in Understanding Man and Society*. 1995 ISBN 0-7923-3627-5
22. A. Camacho: *Division of Labor, Variability, Coordination, and the Theory of Firms and Markets*. 1996 ISBN 0-7923-3832-4
23. R. Hegselmann, U. Mueller and K.G. Troitzsch (eds.): *Modelling and Simulation in the Social Sciences from the Philosophy of Science Point of View*. 1996 ISBN 0-7923-4125-2

24. J. Nida-Rümelin: *Economic Rationality and Practical Reason*. 1997 ISBN 0-7923-4493-6
25. G. Barbiroli: *The Dynamics of Technology*. A Methodological Framework for Techno-Economic Analyses. 1997 ISBN 0-7923-4756-0
26. O. Weinberger: *Alternative Action Theory*. Simultaneously a Critique of Georg Henrik von Wright's Practical Philosophy. 1998 ISBN 0-7923-5184-3
27. A. Carsetti (ed.): *Functional Models of Cognition*. Self-Organizing Dynamics and Semantic Structures in Cognitive Systems. 2000 ISBN 0-7923-6072-9
28. J. Nida-Rümelin and W. Spohn (eds.): *Rationality, Rules, and Structure*. 2000 ISBN 0-7923-6326-4
29. J. Klüver: *The Dynamics and Evolution of Social Systems*. New Foundations of a Mathematical Sociology. 2000 ISBN 0-7923-6443-0
30. V.A. Lefebvre: *Algebra of Conscience*. Revised Edition with a Second Part with a new Foreword by Anatol Rapoport. 2001 ISBN 0-7923-7121-6
31. J. Götschl (ed.): *Evolution and Progress in Democracies*. Towards New Foundations of a Knowledge Society. 2001 ISBN 1-4020-0063-4
32. N.J. Saam and B. Schmidt (eds.): *Cooperative Agents*. Applications in the Social Sciences. 2001 ISBN 1-4020-0190-8
33. B. Verbeek: *Instrumental Rationality and Moral Philosophy*. An Essay on the Virtues of Cooperation. 2002 ISBN 1-4020-0639-X
34. J. Klüver: *An Essay Concerning Sociocultural Evolution*. Theoretical Principles and Mathematical Models. 2002 ISBN 1-4020-0750-7
35. T.J. Fararo and K. Kosaka: *Generating Images of Stratification*. A Formal Theory. 2003 ISBN 1-4020-1500-3
36. O. Dahlbäck: *Analyzing Rational Crime - Models and Methods*. 2003 ISBN 1-4020-1657-3
37. E. Oeser: *Evolution and Constitution*. The Evolutionary Selfconstruction of Law. 2003 ISBN 1-4020-1784-7

MIX
Papier aus verantwortungsvollen Quellen
Paper from responsible sources
FSC® C105338

If you have any concerns about our products,
you can contact us on
ProductSafety@springernature.com

In case Publisher is established outside the EU,
the EU authorized representative is:
Springer Nature Customer Service Center GmbH
Europaplatz 3, 69115 Heidelberg, Germany

Printed by Libri Plureos GmbH
in Hamburg, Germany